# BIOLOGY

## FOR LEAVING CERTIFICATE

ORDINARY LEVEL

**TIM O'MEARA**

educate.ie

*PUBLISHED BY:*
**Educate.ie**
Walsh Educational Books Ltd
Castleisland, Co. Kerry, Ireland
**www.educate.ie**

*EDITOR:*
Susan Power

*DESIGN:*
Kieran O'Donoghue

*PRINTED AND BOUND BY:*
Walsh Colour Print, Castleisland



**ISBN: 978-1-907772-59-7**

# Contents



Please keep your workbook tidy.
Draw all diagrams with a pencil

# Unit 1

# Biology – The Study of Life

# Scientific Method

1. Explain the term 'scientific method'……………………………………………………………

……………………………………………………………………………………………………

……………………………………………………………………………………………………

2. What is a hypothesis? ……………………………………………………………………………

3. What is a theory? ……………………………………………………………………………

4. Scientists investigate scientific questions by doing ……………………………………………

5. Why do scientists carry out experiments? ………………………………………………………

6. Fill in the blank spaces to show the stages involved in the scientific method.

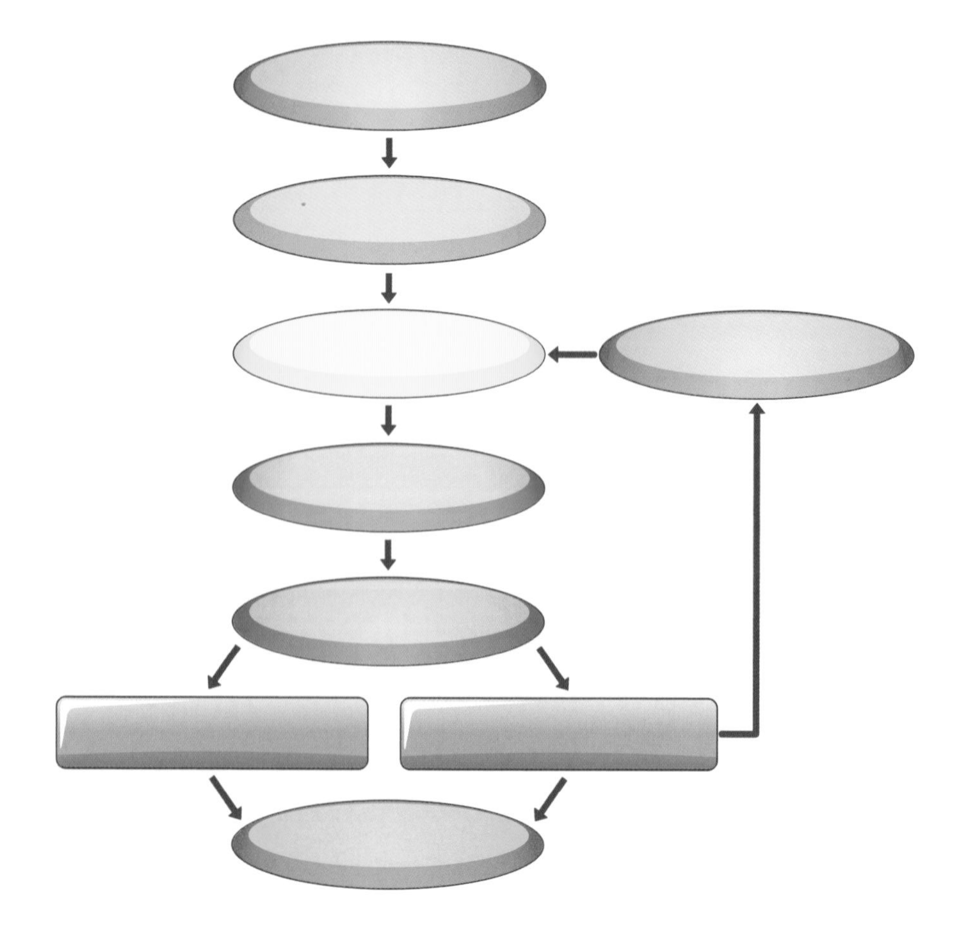

7. Name four steps when planning an experiment ........................................

..........................................................................................

..........................................................................................

..........................................................................................

..........................................................................................

8. Explain the following in relation to carrying out scientific experiments:

a) Having a control ........................................

..........................................................................................

b) Large sample size ........................................

..........................................................................................

c) Double-blind testing ........................................

..........................................................................................

d) Participants selected at random ........................................

..........................................................................................

e) Only one factor changed at one time ........................................

..........................................................................................

f) Repeating the experiment ........................................

..........................................................................................

# Unit 1

# Life: Characteristics of Living Things

1. What is meant by the term 'life'? ..............................

..............................

2. What is meant by the term 'continuity of life'? ..............................

..............................

3. An organism is another name for a ..............................

4. List three characteristics of life ..............................

..............................

5. Match the word in column 1 with the correct explanation in column 2.

| Column 1 | Column 2 |
|---|---|
| 1. Organisation | A. Increase their number. |
| 2. Nutrition | B. The release of energy from food. |
| 3. Respiration | C. Cells form tissues, tissues form organs, organs form systems. |
| 4. Excretion | D. Need food to give them energy. |
| 5. Response | E. Getting rid of waste. |
| 6. Reproduction | F. Respond to changes in their surroundings. |

1 ........ 2 ........ 3 ........ 4 ........ 5 ........ 6 ........

6. What is meant by the term 'tissue'? ..............................

..............................

7. How do living things increase in number? ..............................

..............................

8. Why do cells specialise in multi-celled organisms? ..............................

..............................

9. What is meant by the term 'metabolism'? ..............................

..............................

10. A chemical reaction that builds up a complex molecule from small simple molecules is called ................

.......................................................................................................................................................

Give one example of this type of reaction in biology ..........................................................................

11. Catabolism is what type of chemical reaction? .........................................................................................

.......................................................................................................................................................

Give one example of this type of reaction in living things ...................................................................

12. Name an anabolic process carried out by plants ...............................................................................

13. The enzyme amylase breaks down starch into a simple sugar ...........................................................

This is an example of a ....................................................................................................... reaction.

Unit 1

# Food: The Chemicals of Life

1. What is a biomolecule? ......................................................

......................................................

2. Name four major groups of biomolecules:

   a) ......................................................

   b) ......................................................

   c) ......................................................

   d) ......................................................

3. Living things need food for ......................................................

......................................................

......................................................

4. What is the main function of sugars in the diet? ......................................................

5. Carbohydrates contain the elements ......................................................

   These elements combine together in the ratio ......................................................

6. Name (a) three types of carbohydrate and (b) give one example of each.

   Name.............................................. Example ..............................................

   Name.............................................. Example ..............................................

   Name.............................................. Example ..............................................

7. Give a metabolic use of a fat ......................................................

......................................................

8. Give a structural use of a fat ......................................................

......................................................

9. Give a metabolic use of a carbohydrate ......................................................

......................................................

10. Give a structural use of a carbohydrate ..............................................................................................................

.............................................................................................................................................................

11. Starch is an example of what type of carbohydrate? ..........................................................................

12. Name a polysaccharide found in the cell wall of plant cells ..................................................................

13. Give one metabolic use for lipids in the diet ..........................................................................................

14. Name a reducing sugar ...........................................................................................................................

15. Fat is made up of .......................................................... and ..........................................................

16. Proteins contain the elements ....................................................................................................................

17. Give a metabolic use of protein ..................................................................................................................

18. Give a structural use of a protein ..............................................................................................................

19. Enzymes are made of .................................................................................................................................

20. Proteins are made of long chains of ..........................................................................................................

21. Vitamins are ..................................................................................................................................................

22. a) Name a fat-soluble vitamin ....................................................................................................................

    b) Give a good source of the vitamin you named ....................................................................................

    c) Name the deficiency disease if a person is lacking that vitamin ..........................................................

    d) Give one function of the vitamin you named ........................................................................................

23. Name two minerals needed by the body. Give one use the body makes of the minerals you named.

    Mineral ...................................................... Use ......................................................................

    Mineral ...................................................... Use ......................................................................

24. Fill in the table below.

| **Food** | **Reagent (chemicals used)** | **Positive result** |
|---|---|---|
| Reducing sugar | | |
| Starch | | |
| Fat | | |
| Protein | | |

25. Cellulose is a polysaccharide. Explain the term 'polysaccharide' ..................

..................

Name a polysaccharide other than cellulose ..................

Where precisely in a plant cell would you expect to find cellulose? ..................

26. Name the chemical elements present in carbohydrate ..................

Which two of these elements always occur in a 2:1 ratio? ..................

..................

27. Name a structural carbohydrate ..................

Give a use of carbohydrates other than a structural one ..................

28. Name a chemical element always present in proteins but not in carbohydrates ..................

29. Give one reason why the body needs water ..................

..................

30. Give one way in which water is lost from the body ..................

31. The make-up of a colourless sports drink is to be investigated. Use your knowledge of food testing to answer the following:

a) Name the chemical used to test the sports drink for glucose (reducing sugar).

..................

If glucose is in the drink, what colour change would you expect to see? In your answer give the colour at the start and final colour of the test solution.

Colour at start ..................

Final colour ..................

Is heat necessary for this test? ..................

b) Name the chemicals used to test the sports drink for the presence of protein. If protein is present in the drink, what colour change would you expect to see? In your answer give the colour at the start and final colour of the test solution.

Colour at start ..................

Final colour ..................

Is heat necessary for this test? ..................

# Unit 1

# Ecology

1. Define the following terms:

   Biosphere ..............................

   Habitat ..............................

   Producer ..............................

   Consumer ..............................

   Herbivore ..............................

   Carnivore ..............................

   Omnivore ..............................

   Decomposers ..............................

   Detritus feeder ..............................

   Food chain ..............................

   Food web ..............................

   Pyramid of numbers ..............................

   Niche ..............................

2. Study the following diagram and then answer the questions.

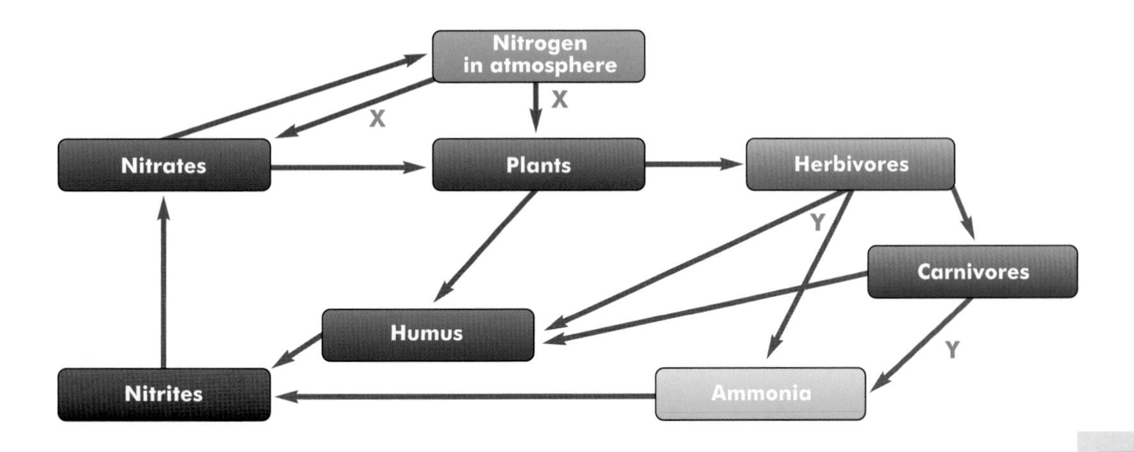

Name the cycle outlined in the diagram ..........

Name an important group of biomolecules that plants make from nitrates ..........

What is happening at any of the processes marked Y? ..........

..........

X shows the change of nitrogen gas to useful products. What is the name given to this process?

..........

Name a group of organisms that carry out X ..........

Why is nitrogen needed by living organisms? ..........

..........

Why is it not advised to add nitrates to the soil in very wet weather? ..........

..........

Why does growing clover make the soil more fertile? ..........

3. Explain the terms:

   Adaptation ..........

   Pollution ..........

4. Give two causes of water pollution.

   a) ..........

   b) ..........

5. Give two causes of air pollution.

   a) ..........

   b) ..........

6. Conservation is ..........

   ..........

   ..........

7. Name two abiotic factors that influence a habitat.

   a) ..........

   b) ..........

8. What is the difference between a quantitative and a qualitative study? ..........

..........

..........

9. Name two ways to sample plants in a habitat ..........

..........

..........

10. Name the animal sampling instruments pictured below.

.......... .......... ..........

.......... .......... .......... ..........

Select two of the instruments from the diagrams above and outline how they are used.

Instrument .......... Use ..........

Instrument .......... Use ..........

11. The diagram shows a food web. Use the web to answer the following questions:

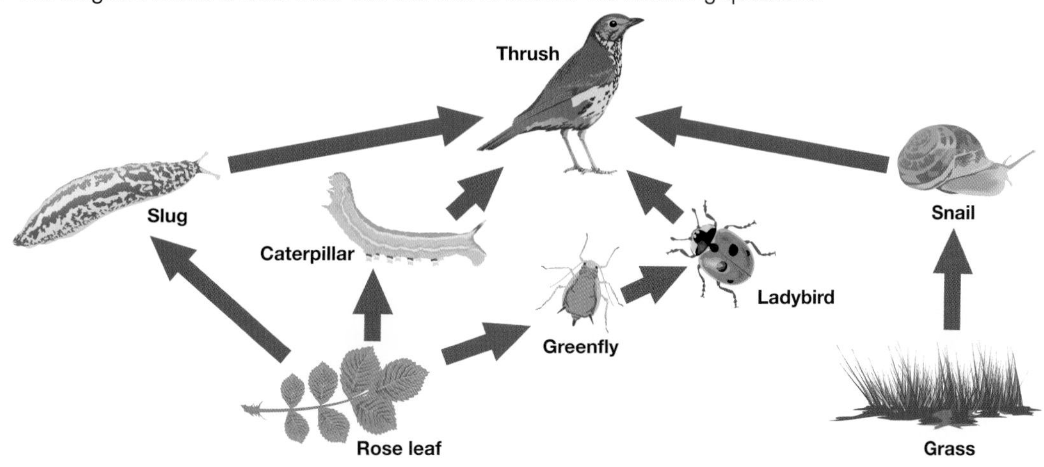

Name two producers ..........................................

Name a primary consumer ..........................................

Name a secondary consumer ..........................................

Name a herbivore ..........................................

Name a carnivore ..........................................

What is the difference between a food chain and a food web? ..........................................

..........................................

12. An animal that eats both plants and animals is called ..........................................
13. What is the main source of energy for the earth's ecosystem? ..........................................
14. What does the pyramid of numbers show? ..........................................
15. The diagram shows a pyramid of numbers. Use the diagram to answer the questions below.

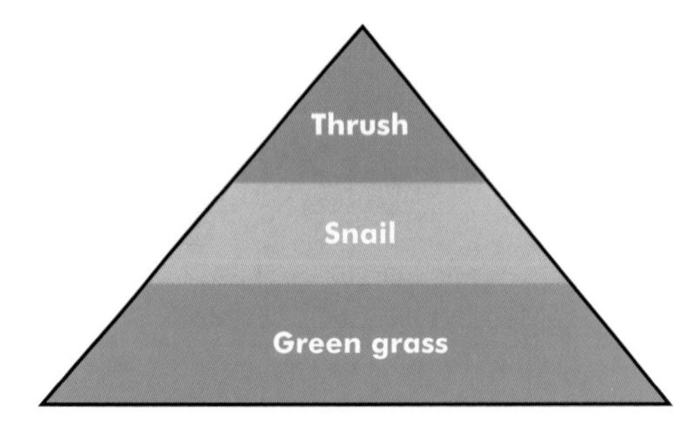


Write out the food chain represented by the pyramid ..........................................

..........................................

Name the organism in trophic level three ..........................................

The largest amount of energy is at which trophic level? ..........................................

Name the primary consumer ..........................................

The producer in the pyramid is ..........................................

16. The following shows a food chain.

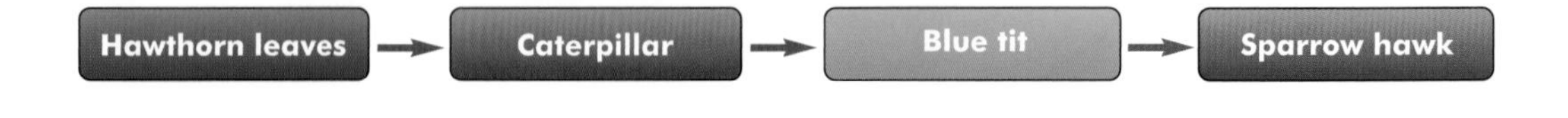


The primary consumer in this food chain is ..........................................

If the number of sparrow hawks increases, the number of blue tits may ..................

In this food chain the hawthorn leaves represent the ..................

Name a carnivore from this food chain ..................

The number of trophic (feeding) levels in this food chain is limited by the small transfer of

..................from one level to the next.

Suggest two ways energy is lost from one trophic level to the next.

a)..................

b)..................

17. Answer the following questions with reference to an ecosystem you have studied.

Name the ecosystem ..................

Name two habitats from the ecosystem ..................

Name an animal that is present in one of these habitats and describe one way in which it is adapted to that habitat.

Name ..................

Adaptation..................

Describe how you collected a named animal ..................

..................

..................

..................

18. Describe how you carried out a quantitative survey of a named plant found in the ecosystem.

..................

..................

..................

..................

..................

..................

..................

19. Give an example of pollution which may result from domestic (household) or industrial or agricultural activity.

    Name of source ..........

    Example of pollution ..........

20. Suggest two ways to control pollution.

    a) ..........

    b) ..........

21. Complete the following sentences by adding the correct term from the list below.

    **habitat, predator, biosphere, niche, ecosystem**

    Kills and eats other animals ..........

    All parts of the earth and its atmosphere, where life exists ..........

    A community of organisms and their environment ..........

    The role of an organism in an ecosystem ..........

    Place where an organism lives ..........

22. The diagram shows a food web. Use the diagram to answer the questions.

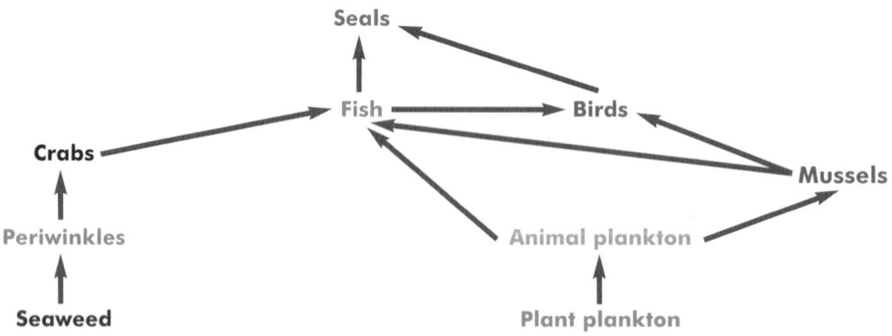

    Name a producer ..........

    What does the animal plankton feed on? ..........

    What feeds on the animal plankton? ..........

    Why are periwinkles referred to as primary consumers? ..........

    Starting with a producer, complete a food chain with four trophic (feeding) levels, naming each organism involved.

    ..........

23. Describe one method of waste management with reference to agriculture, fisheries or forestry.

..........

..........

..........

24. Suggest two ways of minimising waste ..........

..........

25. Name two ecosystems found in Ireland ..........

26. What role do micro-organisms have in the environment? ..........

..........

27. Give one use for a tullgren funnel ..........

28. What is a beating tray used for in a habitat study? ..........

..........

29. The diagram shows a pyramid of numbers for an ecosystem. Answer the following questions:

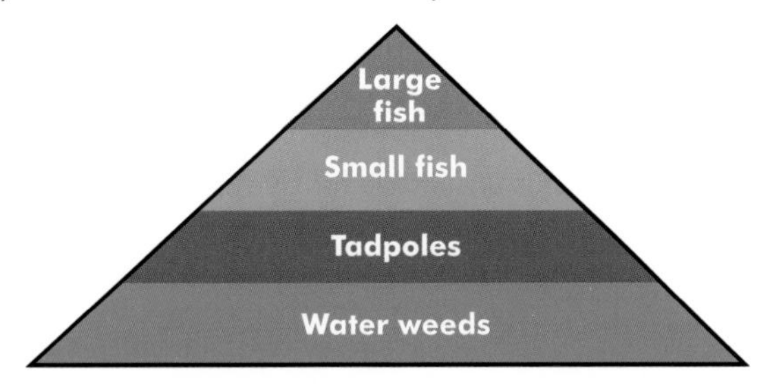


Which organism is a primary consumer? ..........

The largest amount of energy in the pyramid is found in the ..........

Which organism in the pyramid is the producer? ..........

Which organism in the pyramid is the secondary consumer? ..........

An organism which makes its own food is called a ..........

An organism that eats another organism is called a ..........

The place where an organism lives is called its ..........

The primary source of energy in an ecosystem is the ..........

30. The parts of the earth and atmosphere in which life is found is called the ..........

31. Name an ecosystem you have studied ..........

Name three plants that are normally present in this ecosystem.

a) ..........

b) ..........

c) ..........

32. Distinguish between biotic and abiotic factors.

Biotic ..........

Abiotic ..........

33. An <u>edaphic</u> factor is an example of an abiotic factor. Explain the underlined term ..........

..........

34. Name the piece of equipment shown below which is used in a quantitative study of an ecosystem.

50cm
50cm

..........

Why is the above piece of apparatus unsuitable for studying most animal populations? ..........

..........

Suggest a plant that would not be suitable to survey using the above apparatus ..........

..........

Outline how this piece of apparatus is used for studying plant populations ..........

..........

..........

..........

How did you present your results? ..........

State one possible source of error in a survey of an ecosystem ..........

..........

NOTES

# Unit 2

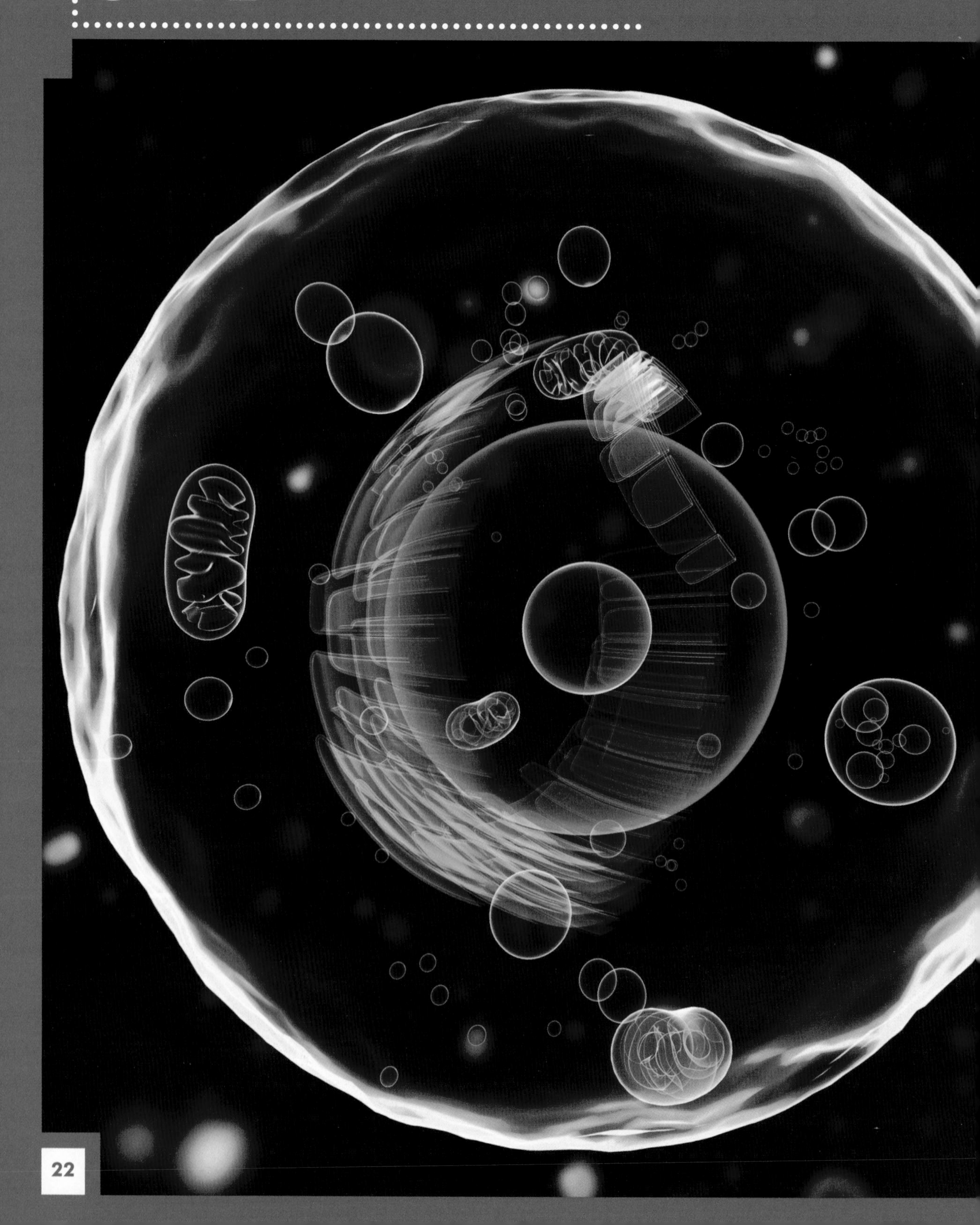

# The Cell

# Microscope

1. Give a use of a microscope ..................................................

..................................................

2. Name the labelled parts of the microscope.

A..................................................

B..................................................

C..................................................

D..................................................

E..................................................

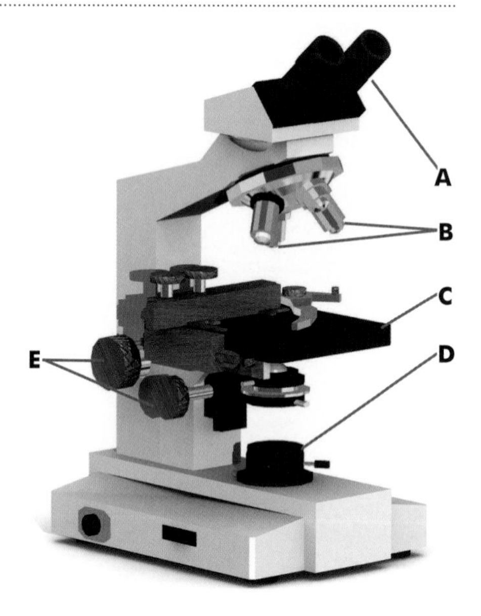


3. When looking at the epidermis of an onion some of the following procedures were followed.

Explain the purpose of each procedure:

Using a very thin section ..................................................

..................................................

Covering the piece of epidermis with a cover slip ..................................................

..................................................

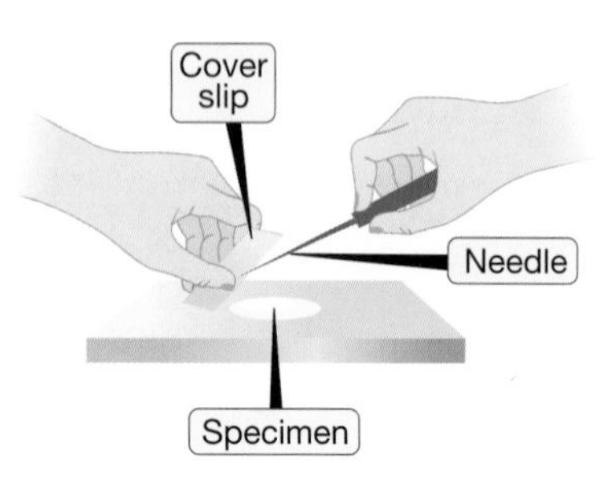

Why it is necessary to lower the cover slip as shown in the diagram? ..............................

..............................

..............................

Adding iodine to the slide ..............................

..............................

Using the low power objective lens first ..............................

..............................

4. When looking at an animal cell under a microscope, explain the following:

Where did you get the cells from? ..............................

What stain did you use? ..............................

Why is a stain used? ..............................

..............................

..............................

Draw a diagram of the result.

# Unit 2

# Cells

1. Living things are made up of living units called ..................................................

2. Name the parts of a plant and animal cell labelled A to F as seen using a microscope.

Animal cell | Plant cell

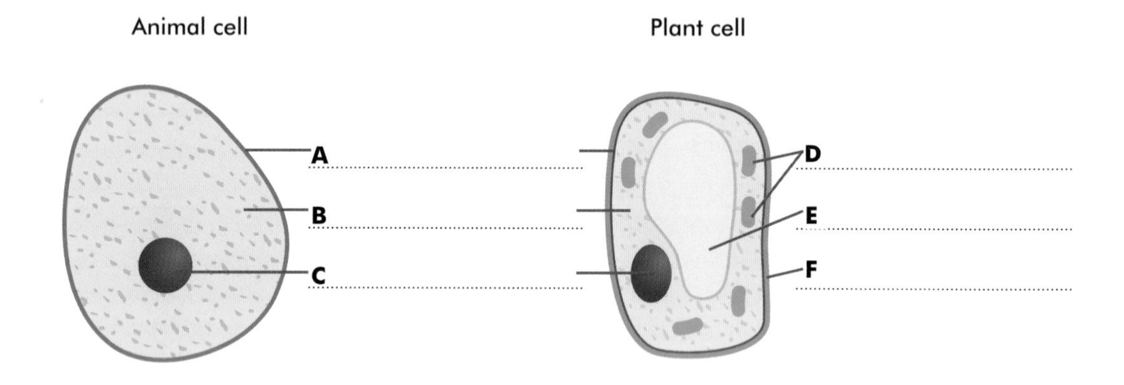


3. Name the parts of an animal cell as seen under an electron microscope.

A ..................................................

B ..................................................

C ..................................................

D ..................................................

E ..................................................

F ..................................................

G ..................................................

H ..................................................

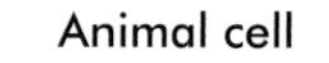


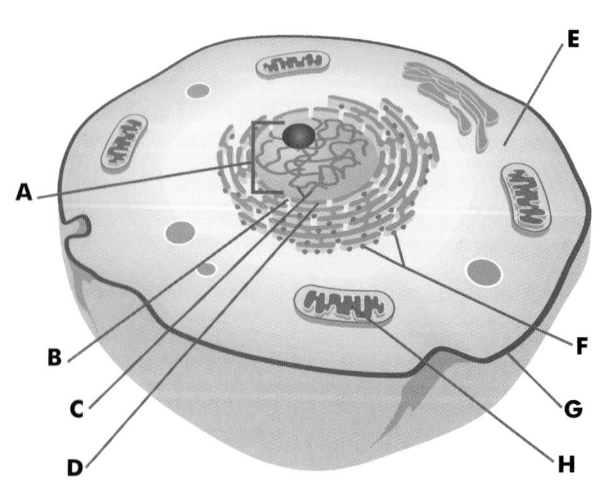


4. Give two functions of a cell membrane.

a) ..................................................................................................

b) ..................................................................................................

5. Name the cell structures A and B.

A

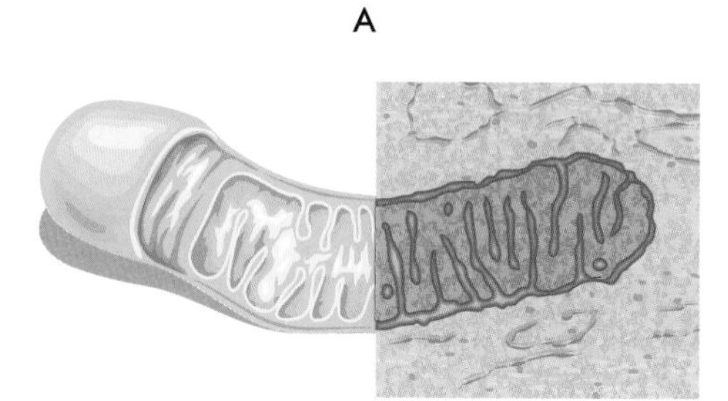

B

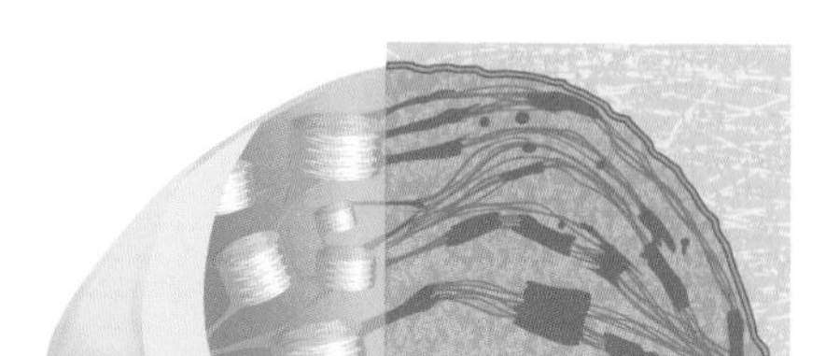

A: ........................................ B: ........................................

What process takes place in A? ........................................

In what cell type is structure B found? ........................................

What process takes place in structure B? ........................................

6. Explain the difference between chromatin and chromosomes ........................................

........................................

........................................

7. Name the part of the cell shown in the diagram ........................................

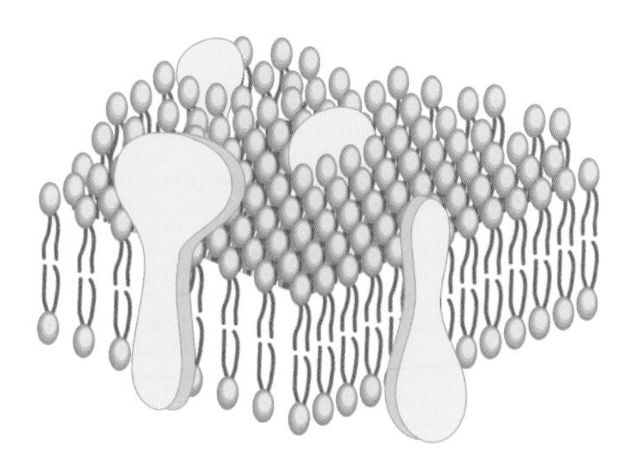

8. Give one function for each of the following cell parts.

Nucleus ........................................

Cytoplasm ........................................

Mitochondrion ........................................

Ribosome ........................................

9. Name the parts of a plant cell as seen under an electron microscope.

A..........

B..........

C..........

D..........

E..........

F..........

G..........

H..........

I..........

Plant cell

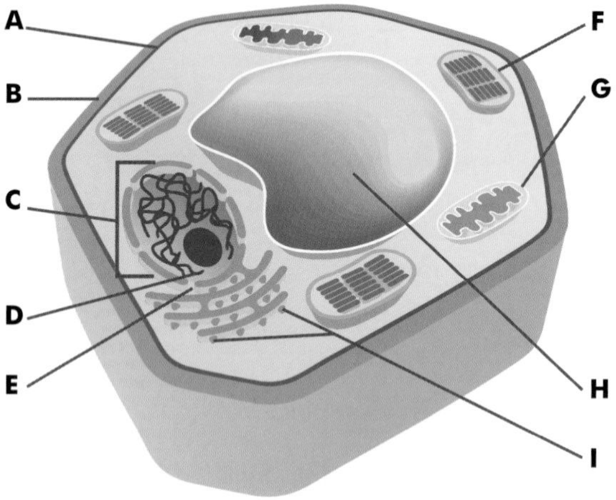

A B C D E F G H I

10. Give one function for each of the following parts of a plant cell.

Cell wall ..........

Chloroplasts ..........

Large vacuole ..........

11. List two differences between a plant and animal cell.

Plant cell

a) .......... b) ..........

Animal cell

a) .......... b) ..........

# Unit 2

## Enzymes

1. What is an enzyme? ..........

..........

2. Enzymes are made of ..........

3. Name two factors that affect the rate of an enzyme.

   a) ..........

   b) ..........

4. What is meant by the optimum temperature for an enzyme? ..........

..........

5. Give two uses made of enzymes in industry.

   a) ..........

   b) ..........

6. Why are enzymes sometimes put in gel beads? ..........

..........

   Name a chemical used to make gel beads ..........

7. The graph below shows the effect of pH on the speed of an enzyme reaction.

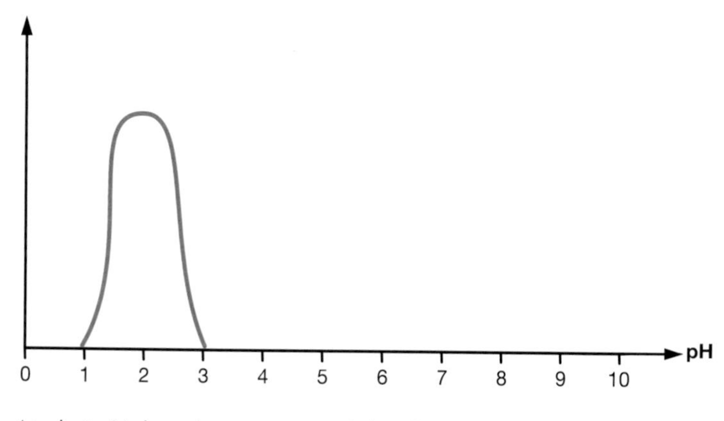


At what pH does the enzyme work best? ..........

What will happen to the speed of the reaction if the pH is increased? ..........

..........

Where in the human body might you find this enzyme? ..........

..........

Draw in the graph the effect of pH on the enzyme in saliva.

8. What is the optimum temperature for enzymes in the human body? ..........

9. What is the name given to the substance an enzyme acts on? ..........

10. Name an enzyme that could be added to washing powder to remove a protein stain.

..........

Give an example of a protein stain ..........

Why should the washing powder containing this enzyme not be used at very high temperatures?

..........

..........

11. Is an enzyme a lipid, a protein or a carbohydrate? ..........

12. Where in a cell are enzymes produced? ..........

13. What is meant by immobilisation of an enzyme? ..........

..........

14. Describe how you immobilised an enzyme in the course of your practical work.

..........

..........

..........

..........

..........

..........

..........

..........

..........

15. Give **one** advantage of bioprocessing using an immobilised enzyme ..........

..........

..........

16. Suggest **one** reason why enzymes are not found in body soap or shampoo ..........

..........

..........

17. To what group of biomolecules do enzymes belong? ..........

Name the small molecules which are the building blocks for these biomolecules.

..........

18. Name **A** and **B** in an enzyme-controlled reaction.

A.......... B..........

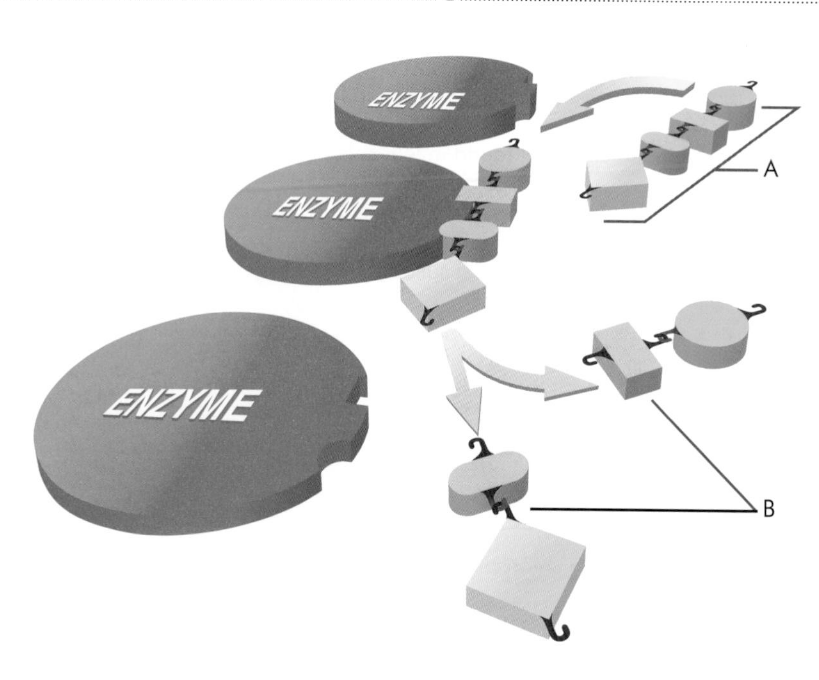

# Unit 2

# Photosynthesis

1. What is meant by nutrition? ..................

2. What is meant by autotrophic nutrition? ..................

3. Explain the term photosynthesis ..................

..................

..................

4. Name the structure in plant cells where photosynthesis takes place ..................

5. Name two raw materials necessary for photosynthesis and state where each is taken in from.

   Material 1 .................. taken in from ..................

   Material 2 .................. taken in from ..................

6. What is the name of the green chemical found in leaf cells? ..................

7. Name the structure in the leaf that lets gases in and out ..................

8. Name two adaptations of the leaf that make it suitable for photosynthesis.

   a) ..................

   b) ..................

9. What are chloroplasts? ..................

..................

10. Name the green chemical found in chloroplasts ..................

11. Where are most chloroplasts found in a plant? ..................

12. Write out the equation for photosynthesis ..................

..................

13. Name two factors that affect the rate of photosynthesis.

   a) ..................

   b) ..................

14. State two ways humans can increase the rate of photosynthesis.

    a) ..........

    b) ..........

15. The energy trapped by chlorophyll is used to split water. Name the three things water is split into.

    a) ..........

    b) ..........

    c) ..........

    Describe what happens to each.

    a) ..........

    b) ..........

    c) ..........

16. Name two products made during photosynthesis and state what happens to each product.

    Name ..........

    What happens to the product? ..........

    Name ..........

    What happens to the product? ..........

17. Suggest two reasons why life on earth might not continue without photosynthesis.

    a) ..........

    b) ..........

# Respiration

1. What is meant by respiration? ......................................................

2. Name the two types of respiration.

   a) ......................................................

   b) ......................................................

3. Write out the equation for respiration ......................................................

   ......................................................

4. What is meant by anaerobic respiration? ......................................................

   ......................................................

5. What are the two products of aerobic respiration?

   a) ......................................................

   b) ......................................................

6. Answer the questions below.

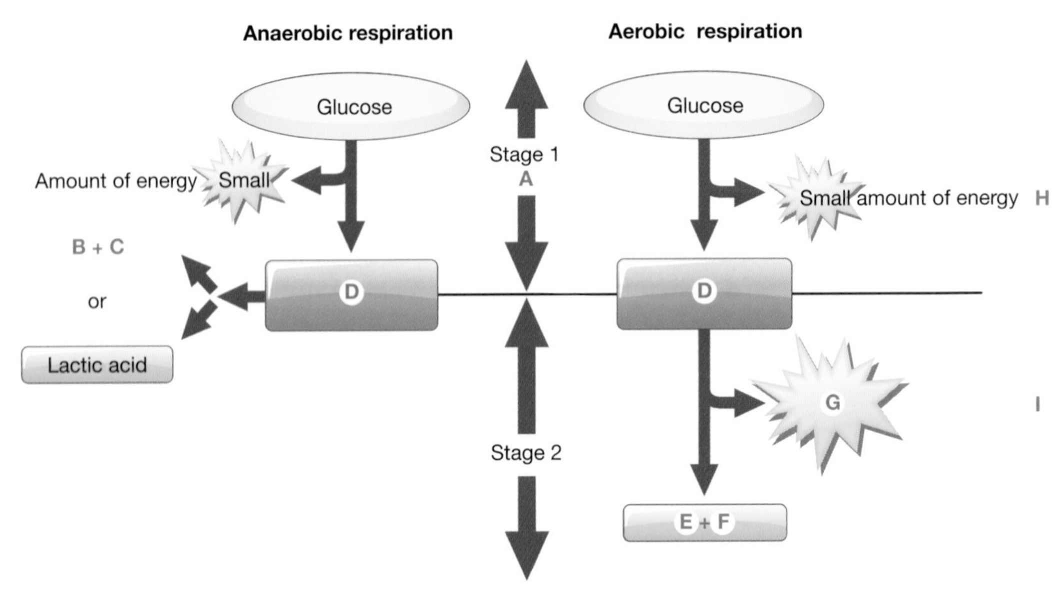


Stage 1 is called A = ......................................................

Name the two products B and C of anaerobic respiration in yeast.

B........................................ C........................................

Stage 1 takes place in H. H = ........................................

Name the product of stage 1. D = ........................................

Stage 2 takes place in I. I = ........................................

The end products of stage 2 are E and F. Name E and F

E........................................ F........................................

7. What is fermentation? ........................................

   Name the industry that uses this type of reaction? ........................................

8. Indicate whether the following are true (T) or false (F) by drawing a circle around T or F.

   Carbon dioxide is produced during respiration ........................................ T F

   Stage 1 of respiration requires oxygen ........................................ T F

   Stage 1 of respiration takes place in the cytoplasm ........................................ T F

   Stage 2 of respiration also takes place in the cytoplasm ........................................ T F

   Some of the energy released in respiration is lost as heat ........................................ T F

   Lactic acid is a product of anaerobic respiration ........................................ T F

9. In stage 1 of respiration, glucose is partly broken down. Where in the cell does this happen?

   ........................................

10. Name the cell component in which stage 2 of respiration takes place ........................................

11. Which stage of respiration releases more energy? ........................................

12. Draw a labelled diagram of the apparatus used to produce alcohol using yeast.

In your investigation it was necessary to exclude air. How was this done? ..........

..........

13. Briefly describe a test to show that alcohol has been produced ..........

..........

..........

The test is called ..........

14. Aerobic respiration takes place in two main stages: stage 1 and stage 2.

Indicate clearly whether each of the following statements refer to stage 1 or to stage 2.

Takes place in the mitochondria ..........

Produces a large amount of energy ..........

Takes place in the cytoplasm ..........

Does not require oxygen ..........

# Unit 2

# Movement through a Cell Membrane

1. Name the ways substances can move in and out of a cell.

   a) ....................

   b) ....................

   c) ....................

2. Define what is meant by the term diffusion ....................

   ....................

3. Name two things that affect the speed of diffusion.

   a) ....................

   b) ....................

4. What is osmosis? ....................

   ....................

5. What is meant by the term semi-permeable or selectively permeable? ....................

   ....................

   ....................

6. Name a structure in the cell that is semi-permeable ....................

7. What is meant by turgor? ....................

   ....................

   ....................

8. What is meant by wilting? ....................

   ....................

9. What is meant by osmoregulation? ....................

   ....................

10. How does a single-celled organism living in fresh water get rid of excess water? ........................................

........................................................................................................................................

11. Draw a diagram of a plant cell when it is put into (a) fresh water and (b) concentrated salt solution.

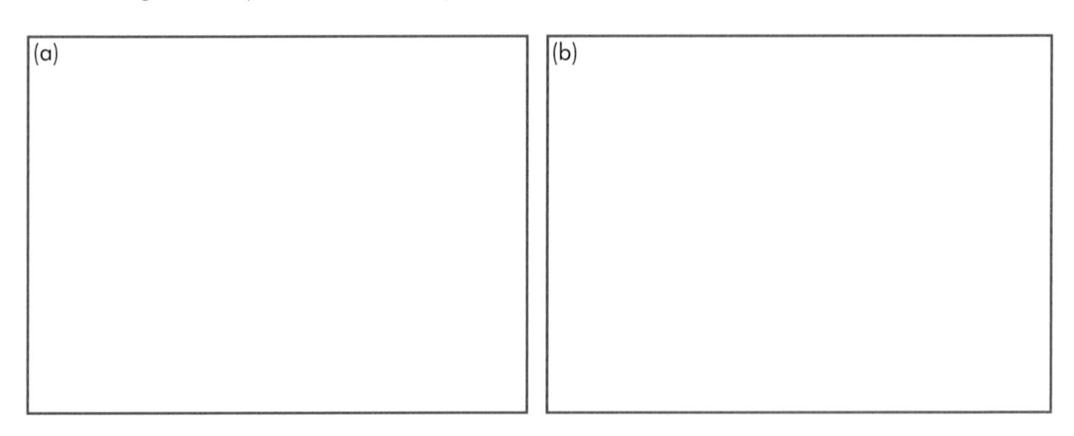


12. A method of food preservation uses osmosis to make food last longer.

Name the method ........................................................................................................

How does it work? ........................................................................................................

........................................................................................................................................

........................................................................................................................................

Give an example of this method of preservation ........................................................

Name two common chemicals used to preserve food using this method.

a) ........................................................................................................................................

b) ........................................................................................................................................

13. The diagram shows two lengths of dialysis tubing. **A** is filled with a sucrose solution and **B** is filled with distilled water. Both are put into beakers of distilled water. Explain what happens in each visking tube and the reason why.

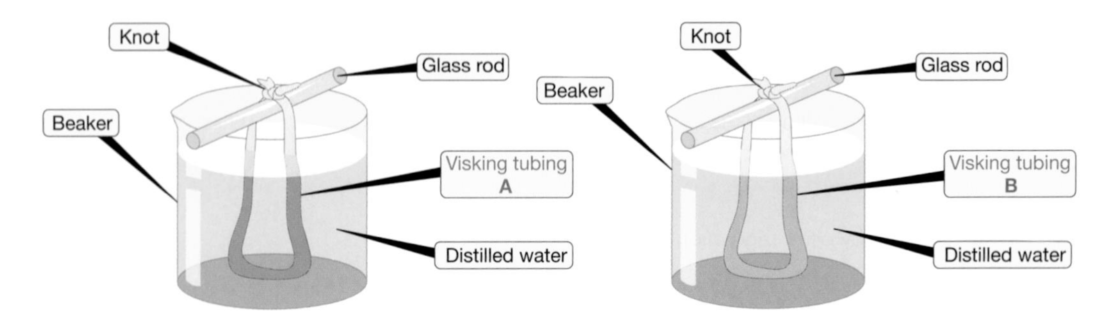

Visking tubing A ..............................................................................

Reason ..............................................................................

..............................................................................

..............................................................................

Visking tubing B ..............................................................................

Reason ..............................................................................

..............................................................................

..............................................................................

14. Which cell, A or B, was in a concentrated salt solution? ..............................................................................

What term is used to describe the condition of cell A? ..............................................................................

Name parts C = ..............................................................................

D = ..............................................................................

E = ..............................................................................

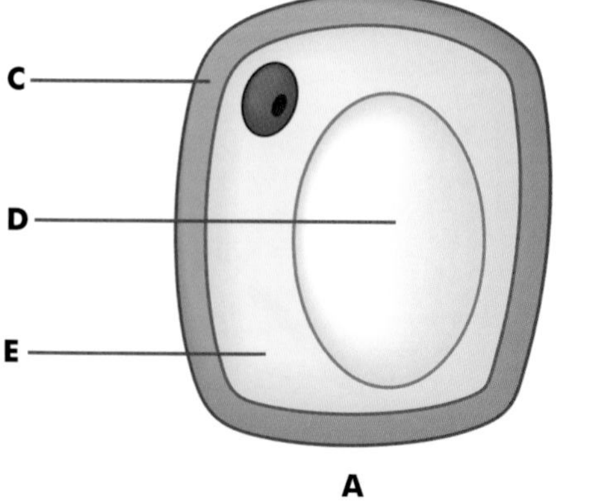


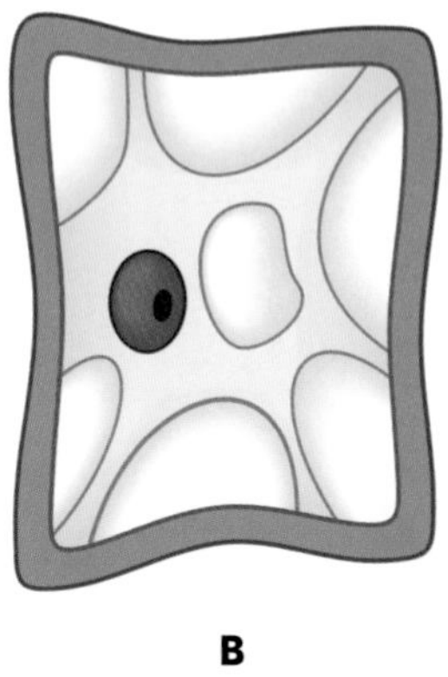

# Unit 2

# Cell Continuity

1. What is meant by cell continuity? ..........

2. A new cell goes through three stages. Name the stages ..........

   ..........

3. What is meant by cell cycle? ..........

4. What is meant by the term interphase? ..........

   ..........

5. Name the two types of cell division.

   a) ..........

   b) ..........

6. Write the name of the type of cell division shown in diagrams A and B.

   A.......... B..........

Parent cell

Parent cell

In the diagrams above, fill in the number of chromosomes present in the new cells in each type of cell division.

7. Define the following terms:

   Chromosome ..........

   Haploid ..........

   Diploid ..........

8. Name the stage in the cell cycle shown in the diagram.

..........................................................................................

State two things that happen during this stage.

a) ..........................................................................................

b) ..........................................................................................

9. The diagrams below show the stages of .............................................. but not in the correct order.

List the stages in the correct order.

A B C D

..........................................................................................

10. Outline the importance of mitosis in:

a) single-celled organism ..........................................................................................

b) multi-celled organism ..........................................................................................

11. What is cancer? ..........................................................................................

12. Give a possible cause of cancer ..........................................................................................

..........................................................................................

13. State two ways to treat cancer.

a) ..........................................................................................

b) ..........................................................................................

14. Give two causes that increase the risk of cancer.

a) ..........................................................................................

b) ..........................................................................................

15. What is a carcinogen? ..........................................................................................

Name a carcinogen ..........................................................................................

# Unit 2

## Cell Organisation

1. Define the following terms:

   Tissue ..........

   Organ ..........

   System ..........

2. Name two types of plant tissue.

   a) ..........

   b) ..........

3. Name two types of animal tissue.

   a) ..........

   b) ..........

4. Name one plant and one animal organ.

   Plant organ ..........

   Animal organ ..........

5. Name one plant and one animal system.

   Plant system ..........

   Animal system ..........

6. What are monoclonal antibodies? ..........

   ..........

   ..........

   ..........

   Give one use of monoclonal antibodies ..........

   ..........

   ..........

7. What is meant by tissue culture? ..............................

..............................

Give one example of tissue culture ..............................

8. Describe stages A, B and C involved in micropropagation.

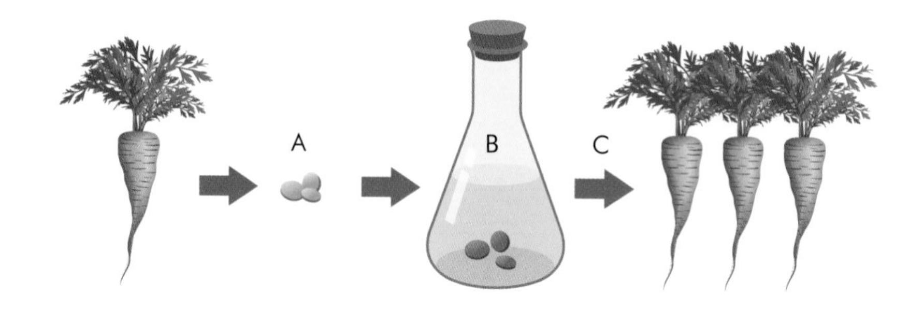


A ..............................

B ..............................

C ..............................

# Unit 2

# Variations

1. What is meant by classification? ........................................

........................................

2. In classification what is meant by a species? ........................................

........................................

3. What is meant by the term variation? ........................................

........................................

4. Name two types of variation.

a) ........................................

b) ........................................

5. Pick out an inherited variation and a non-inherited variation from the diagrams opposite.

Inherited variation ........................................

Non-inherited variation ........................................

What is another name for a non-inherited variation? ........................................

6. Using the list below, pick out two inherited and two non-inherited variations.

**blue eyes, ability to speak French, ability to read, red hair, scar on cheek, curly hair**

Non-inherited ........................................

Inherited ........................................

7. What is an acquired variation? ........................................

8. What is meant by inherited variation? ........................................

Give one example in humans ........................................

# Unit 2

# Heredity and Gene Expression

1. Explain the following terms:

   Heredity ..........

   Chromosome ..........

   Gene ..........

2. Where are chromosomes found in a cell? ..........

3. Chromosomes are made up of .......... + ..........

4. Fill in the names in the diagram.

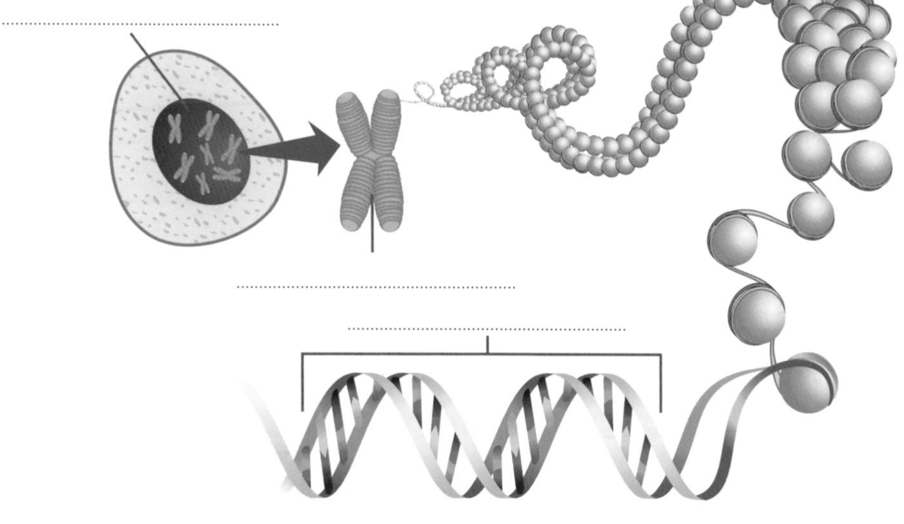

   ..........

   ..........

   ..........

5. What is gene expression? ..........

   ..........

6. What is meant by the coding part of DNA? ..........

   ..........

7. What is the function of DNA? ..........

8. The DNA molecule consists of .......... strands joined by complementary..........

9. Name the four different bases found in DNA ...........................................

...........................................

10. Fill in the names of the missing bases in the diagram of DNA.



...........................................

...........................................

11. How is information carried on the DNA molecule? ...........................................

...........................................

12. Outline how DNA replicates itself ...........................................

...........................................

...........................................

...........................................

...........................................

13. Why is it necessary for DNA to make a copy of itself? ...........................................

...........................................

...........................................

14. Name the nitrogenous bases in DNA whose first letters are A and C.

A........................................... C...........................................

15. Where in the cell would you expect to find most DNA? ...........................................

16. DNA contains the instructions needed to make protein.

These instructions are called the ...........................................code.

17. List two differences between DNA and RNA.

a) ...........................................

b) ...........................................

18. Give one use in the cell for messenger RNA (mRNA) ...........................................

...........................................

19. What is a DNA profile? ...................................................................................

.................................................................................................................

20. Give two uses of a DNA profile.

a) .............................................................................................................

b) .............................................................................................................

21. The stages involved in DNA profiling are listed below, but they are not in the correct order.

A = The DNA is cut into small pieces.

B = The pattern formed by the DNA is examined.

C = The cells are broken down.

D = The DNA pieces are separated.

Rewrite the letters so that they are in the correct order ..........................................

What was used to cut the DNA into pieces? ....................................................

On what basis are the DNA pieces separated? ..................................................

How are the DNA pieces separated? ..............................................................

22. What do the letters DNA stand for? ..............................................................

23. What is meant by genetic screening? ............................................................

.................................................................................................................

.................................................................................................................

24. Outline any ethical consideration in relation to genetic screening ...........................

.................................................................................................................

.................................................................................................................

25. Where are proteins made in the cell? ............................................................

26. The process of protein synthesis involves the following stages, but they are not in the correct order.

A= The code is translated.

B = DNA contains the code for making protein.

C = The protein folds into a particular shape.

D = The code is transcribed into mRNA.

E = Amino acids are assembled in the correct order to make the protein.

F = The transcribed code goes to the ribosome.

Rewrite the letters so that they are in the correct order ........................................

27. What is mRNA? ........................................

........................................

28. Why is the code for an amino acid called a triplet? ........................................

29. To which structures in the cell does mRNA carry the code? ........................................

30. Why is the order of the bases in DNA important? ........................................

........................................

31. What is meant by a triplet code? ........................................

........................................

32. The triplet code is transcribed into mRNA. What does this statement mean? ........................................

........................................

........................................

33. Name the plant from which you isolated DNA in your practical studies ........................................

For what precise purpose did you use freezer-cold ethanol (alcohol) in your isolation of DNA?

........................................

........................................

........................................

# Unit 2

# Genetic Crosses

1. Meiosis is necessary in sexual reproduction. True or false? ..........

   Explain your answer ..........

   ..........

2. What is a gamete? ..........

3. Gametes are made as a result of what type of cell division? ..........

4. Each characteristic or trait in an organism is controlled by how many genes? ..........

5. Explain the following terms:

   Phenotype ..........

   ..........

   Genotype ..........

   ..........

   Allele ..........

   ..........

   Homozygous ..........

   ..........

   Heterozygous ..........

   ..........

   Dominant gene ..........

   ..........

   Recessive gene ..........

   ..........

   Monohybrid cross ..........

   ..........

Incomplete dominance ..........

6. Give one example of incomplete dominance in (a) plants and (b) animals.

   a) ..........

   b) ..........

7. What is a chromosome? ..........

   ..........

8. a) The sex of an individual is controlled by ..........

   b) A female has two .......... chromosomes.

   c) A male has one .......... chromosome and one .......... chromosome.

9. What is the haploid number of chromosomes found in the human egg cell and sperm cell? ..........

10. Hair colour in humans is genetically controlled. The allele for brown hair (B) is dominant to the allele for red hair (b).

    For hair colour Seán is heterozygous (Bb) and Máire is homozygous (bb).

    What colour is Seán's hair? ..........

    What colour is Máire's hair? ..........

    In the space below, use a punnet square or any other means to show the following:

    the genotypes of all the gametes that Seán and Máire can produce

    the genotypes of the children that Seán and Máire may have.

    What is the probability that one of their children may have red hair? (Give your answer as a ratio or a percentage).

11. Explain what is meant by the term gene ..............................................................................

..............................................................................................................................................

12. Where in the nucleus would you find genes? ..............................................................................

13. The allele for brown eyes (B) is dominant to the allele for blue eyes (b).

    In the space below, use a punnet square to find the possible genotypes of children whose parents are both heterozygous for brown eyes.

    State the eye colour resulting from each of these genotypes.

14. In some plants, the allele for red petals (R) is incompletely dominant (codominant) to the allele for white petal (r), the phenotype for the heterozygous condition Rr is pink.

    Show the cross between two pink parents.

    Parents Pink X Pink

    Gametes ..............................................................................................................................

    The genotypes of the offspring

| | | |
|---|---|---|
| | | |
| | | |

What are the genotypes of the offspring? ( ) ( ) ( )

What are the phenotypes of the offspring?

If 120 new plants were produced in this cross, how many of them would you expect to have

pink flowers? ..................................................

Explain how you got this answer ..................................................

..................................................

15. Hairy-stemmed tomato plants were crossed with smooth-stemmed tomato plants.
All the next generation (F1) of plants had hairy stems.
Fill in the blank spaces between the brackets.

Genotypes of parents ( ) X ( )

Genotype of gametes produced ( ) ( )

Genotype of offspring ( )

# Mutation and Evolution

1. Explain the terms:

   Variation ..........

   Mutation ..........

2. Name two types of mutation ..........

3. What is evolution? ..........

4. Name the scientist who put forward the theory of evolution.

5. What evidence is used to support the theory of evolution? ..........

6. What is meant by natural selection? ..........

   Give one example of natural selection ..........

7. Variation is essential for natural selection. Mutation can give rise to variation.

   Give two causes of mutation.

   a) ..........

   b) ..........

# Unit 2

# Genetic Engineering

1. Explain what is meant by genetic engineering ....................................................................

....................................................................................................

....................................................................................................

2. What is used to cut DNA? ....................................................................................

3. What is removed from the donor cell during the process of genetic engineering? ..............

4. Name the structure removed from the bacterial cell.

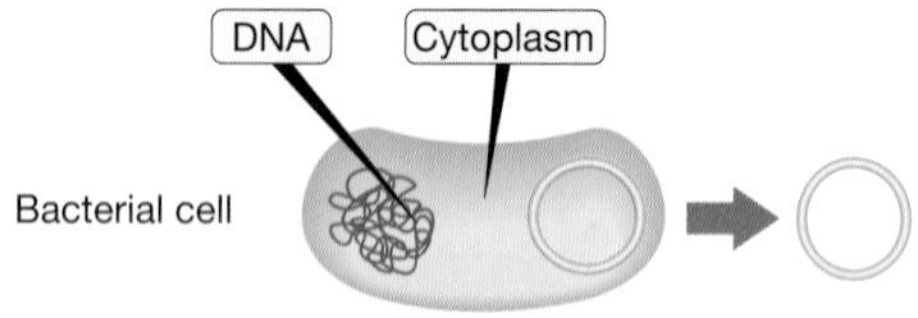

DNA

Cytoplasm

Bacterial cell ..........................................

What happens to the structure after the desired gene has been joined to it? ..........................

....................................................................................................

The bacterial cell then .................................................... making a large number

of cells with the ..............................................................................

5. Name three stages involved in genetic engineering.

a) ..............................................................................................

b) ..............................................................................................

c) ..............................................................................................

6. Give an example of an application of genetic engineering in each of the following cases:

a) A micro-organism ............................................................................

b) An animal ....................................................................................

c) A plant ......................................................................................

## NOTES

# Unit 3

# The Organism

# Variety of Organisms

1. Living things are divided into five kingdoms. Name the five kingdoms and give one feature for each.

Name ..........

Feature ..........

Name ..........

Feature ..........

Name ..........

Feature ..........

Name ..........

Feature ..........

Name ..........

Feature ..........

2. Name the parts of the bacterium labelled A to G.

A ..........

B ..........

C ..........

D ..........

E ..........

F ..........

G ..........



What is the function of part C? ..........

..........

..........

3. Name the three different shapes of bacteria.

   ......................................................

   ......................................................

   ......................................................

4. Under difficult conditions bacteria form ......................................................

5. Bacteria reproduce by ......................................................

6. Name three factors that affect the growth of bacteria.

   a) ......................................................

   b) ......................................................

   c) ......................................................

7. Give two useful effects and two harmful effects of bacteria.

   Useful a) ......................................................

   b) ......................................................

   Harmful a) ......................................................

   b) ......................................................

8. What is a pathogen? ......................................................

   Give one example of a pathogen ......................................................

9. What are antibiotics? ......................................................

   Name an antibiotic ......................................................

10. Why is taking an antibiotic for the common cold not recommended? ......................................................

    ......................................................

11. Give one harmful effect caused by the overuse of antibiotics ......................................................

    ......................................................

12. Outline three safety precautions when working with bacteria in the laboratory.

    a) ......................................................

    ......................................................

    b) ......................................................

..........................................................................................

c)..........................................................................................

..........................................................................................

13. It is important to use sterile apparatus when working with micro-organisms.

What is meant by sterile?..........................................................................

..........................................................................................

How do you sterilise apparatus?..................................................................

..........................................................................................

14. What is meant by the term asepsis?............................................................

..........................................................................................

15. Some bacteria have a layer outside the cell wall. Name this layer and state its function.

Name ..........................................................................................

Function ..........................................................................................

16. A parasite gets its food from a ..........................................................

17. Bacteria form .................................................... if conditions become unfavourable.

18. What is binary fission? ..........................................................................

19. Some bacteria are anaerobic. What does this mean?..........................................

..........................................................................................

20. Give two examples of the economic importance of bacteria.

a)..........................................................................................

b)..........................................................................................

21. Decomposition is important for adding nutrients to the soil. Explain the term decomposition ....................

..........................................................................................

22. Name two groups of micro-organisms in soil responsible for decomposition.

a)..........................................................................................

b)..........................................................................................

# Unit 3

## Kingdom Fungi

1. Give one feature of the Kingdom Fungi ..............................

2. The tiny tubes that make up most fungi are called ..............................

3. What is a mycelium? ..............................

4. Most fungi have cell walls made of ..............................

5. Fungi have no .............................. and so are not able to make their own food.

6. Fungi are heterotrophs. What is meant by the term heterotroph? ..............................

   ..............................

7. What is a parasitic fungus? ..............................

   Give one example ..............................

8. What is a saprophytic fungus? ..............................

   Give one example ..............................

9. List two conditions that help the growth of fungi.

   a) ..............................

   b) ..............................

10. Give two helpful effects and two harmful effects of fungi.

    Helpful a) ..............................

    b) ..............................

    Harmful a) ..............................

    b) ..............................

11. Rhizopus is a saprophytic mould that grows on ..............................

    ..............................

12. What is meant by asexual reproduction? ..............................

    ..............................

13. Yeast cells reproduce asexually by ..........

14. Name the parts labelled A to F in the diagram.

    A..........

    B..........

    C..........

    D..........

    E..........

    F..........

15. Rhizopus has three types of hyphae. Name them.

    a)..........

    b)..........

    c)..........

    Name the hyphae that absorb food..........

16. Rhizopus also reproduces by sexual reproduction. Fill in the missing words in the diagram.

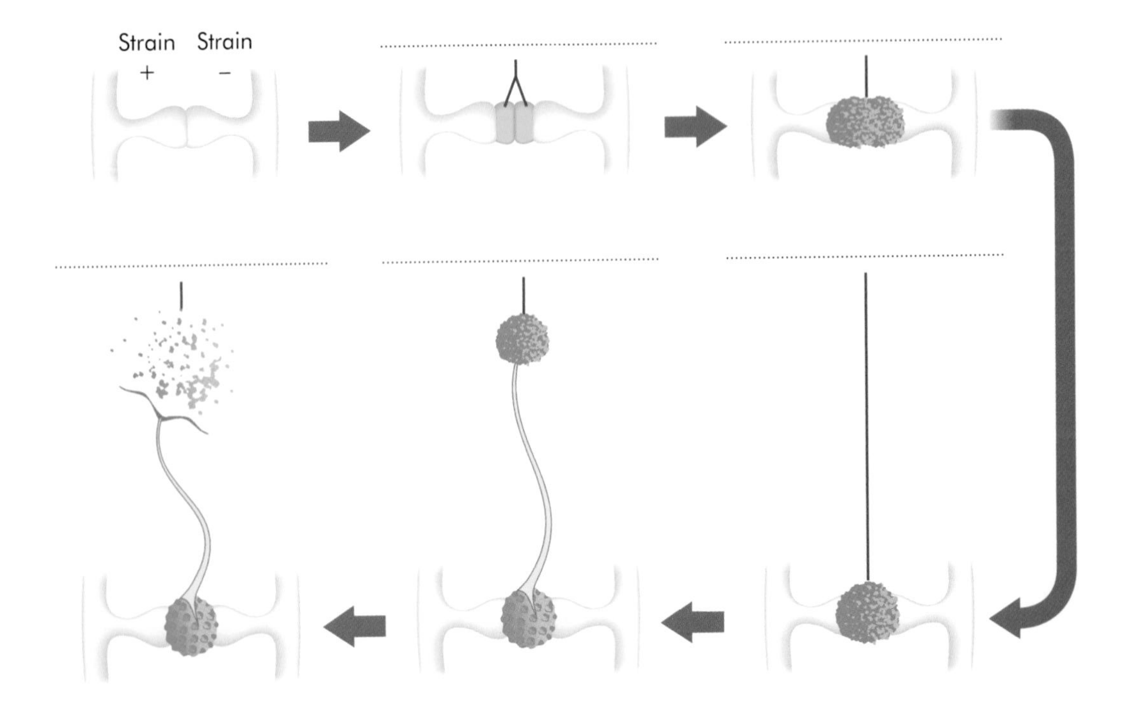

17. Name the parts labelled A to E in the diagram.

**Yeast cell**



A ..............................

B ..............................

C ..............................

D ..............................

E ..............................

**Asexual reproduction in yeast**

Fill in the missing words.

..............................

Nucleus ..............................

..............................

18. What is fermentation? ..............................

..............................

..............................

19. Name an industry that uses fermentation ..............................

20. Complete the following equation

**Sugar + Yeast →** .............................. + ..............................

21. To which kingdom does Rhizopus belong? ..............................

22. Draw a diagram to show the structure of Rhizopus and label three parts.

23. Give an example of a beneficial organism and a harmful organism that belong to the same kingdom as Rhizopus.

    Beneficial ..........

    Harmful ..........

24. One of your practical activities was to prepare alcohol using yeast.

    Name the solution in which you placed the yeast at the start of the activity ..........

    ..........

    Give the temperature at which you then kept the solution ..........

    How did you know that alcohol production had ceased? ..........

    ..........

    Name the test or chemical(s) used to show that alcohol had been produced ..........

    ..........

    Outline any safety precaution ..........

    ..........

# Unit 3

# Kingdom Protista - Amoeba

1. Name an organism that belongs to the Kingdom Protista ..............................................

2. Where does the organism you have named in Q1 live? ..............................................

3. Give one feature of the organism ..............................................

4. Name the parts of the amoeba labelled A to F:

   A..............................................

   B ..............................................

   C..............................................

   D..............................................

   E..............................................

   F..............................................

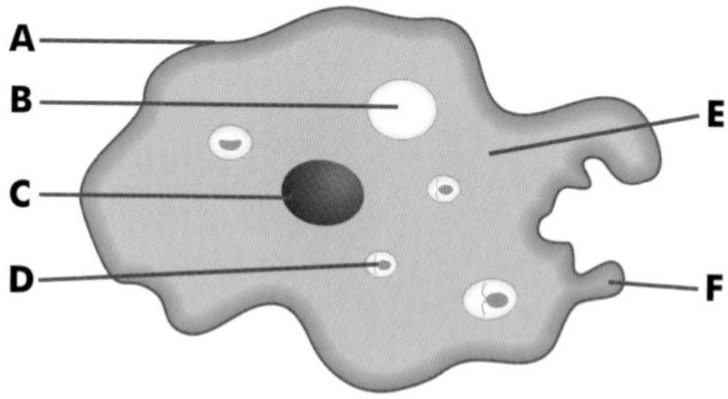

A
B
C
D
E
F

5. Oxygen passes into amoeba by ..............................................

6. Amoeba reproduces asexually by ..............................................

   Asexual reproduction in amoeba involves what type of cell division? ..............................................

   Describe, with the help of diagrams, this type of reproduction in amoeba.

7. What is meant by osmoregulation? ..............................................

   ..............................................

8. Name the structure in amoeba responsible for osmoregulation ..............................................

9. Name a disease caused by amoeba ..............................................

   Give one symptom of the disease ..............................................

# Unit 3

# Plant Kingdom

1. List three features of flowering plants ..........

   a) ..........

   b) ..........

   c) ..........

2. Name the parts of the plant labelled A to K in the diagram.

   A ..........

   B ..........

   C ..........

   D ..........

   E ..........

   F ..........

   G ..........

   H ..........

   I ..........

   J ..........

   K ..........

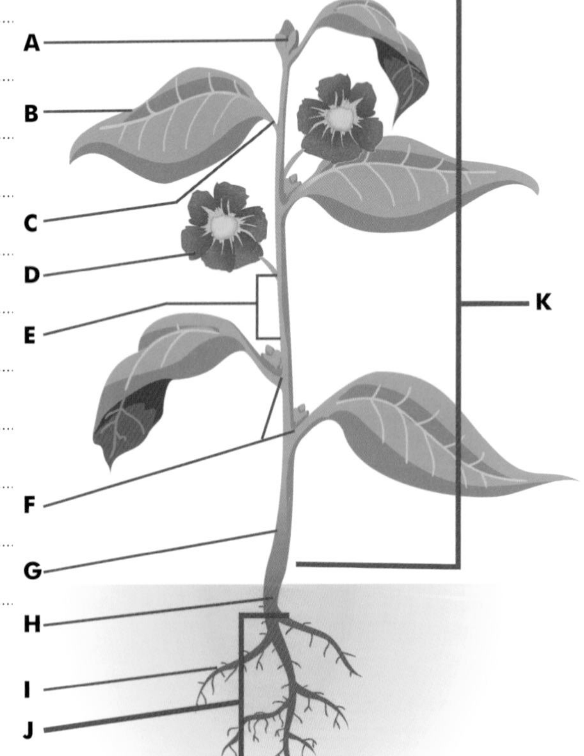


3. Give three functions of the root.

   a) ..........

   b) ..........

   c) ..........

4. Name the two types of root system shown in the diagram. Give one example of each type.

Name:............................................ Name:............................................

Example:............................................ Example:............................................

5. What is an adventitious root? ............................................

............................................

............................................

Give one example of an adventitious root ............................................

6. Give three functions of the stem.

a)............................................

b)............................................

c)............................................

7. Give two functions of the leaf.

a)............................................

b)............................................

8. Name a monocotyledon plant ............................................

9. Name a dicotyledon plant ............................................

10. What is a cotyledon?............................................

............................................

............................................

11. Name the parts of the leaf labelled A to D in the diagram.

A........................................

B........................................

C........................................

D........................................



12. Name the two types of leaf shown and give one example of each.

Name ........................................

Example ........................................

Name ........................................

Example ........................................

13. Give two differences between monocotyledons and dicotyledons.

| Monocotyledon | Dicotyledon |
|---|---|
| a) ........................................ | a) ........................................ |
| ........................................ | ........................................ |
| b) ........................................ | b) ........................................ |
| ........................................ | ........................................ |

14. Describe in detail how you prepared a microscope slide of a transverse section of the stem of a dicotyledonous plant.

........................................

........................................

........................................

........................................

........................................

........................................

15. Give an account of the procedures that you followed in order to view your slide under the microscope.

........................................................................................................................

........................................................................................................................

........................................................................................................................

........................................................................................................................

........................................................................................................................

16. What is meant by meristematic tissue? ..........................................................................

........................................................................................................................

17. Name two places in a plant where meristematic tissue is found.

a) ........................................................................................................................

b) ........................................................................................................................

18. Name the three types of plant tissue shown in the diagrams below.



A ........................................................................................................................



B ........................................................................................................................



C ........................................................................................................................

State two places in a plant where you would find tissue B and C.

B ........................................................................................................................

C ........................................................................................................................

19. What is meant by vascular tissue? ..........................................................................

........................................................................................................................

20. Name two types of vascular tissue in plants.

a) ........................................................................................................................

b) ........................................................................................................................

21. What is the function of xylem tissue? ..............................................................................

..............................................................................................................................

22. Name the tissues marked A and B in the diagram below.

A.................................................................. B ..................................................................

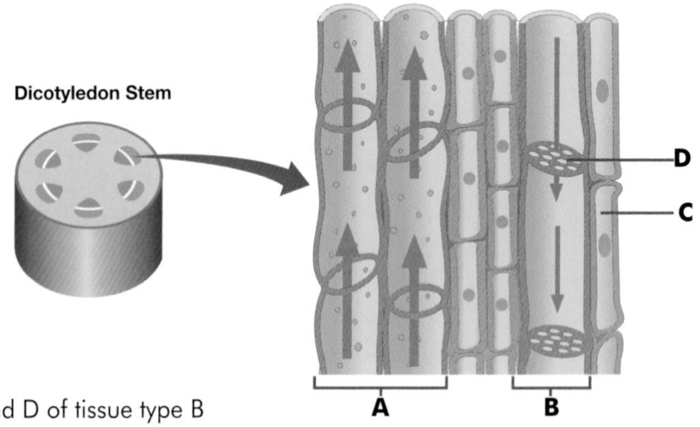

Dicotyledon Stem

D

C

A

B

Name parts C and D of tissue type B

C.................................................................. D..................................................................

What kind of stem is shown in the diagram? ..............................................................................

Explain reasons for your decision ..............................................................................

..............................................................................................................................

Name a material carried in tissue B ..............................................................................

As well as transport, tissue A has another function. What is that function?..............................................

..............................................................................................................................

23. Give one location where mitosis occurs in flowering plants ..............................................................

24. Name the two types of xylem tissue shown in the diagram.

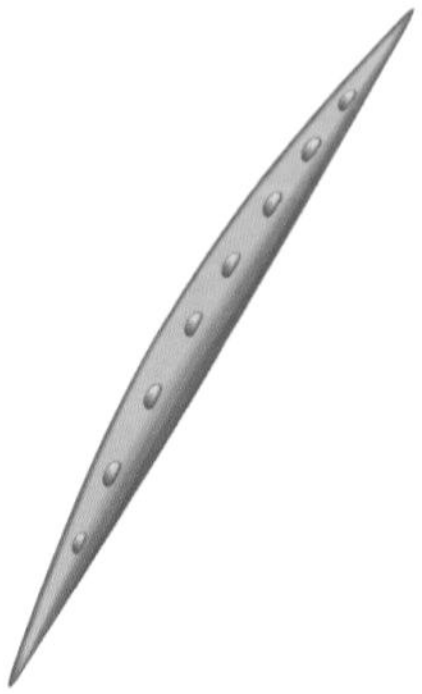

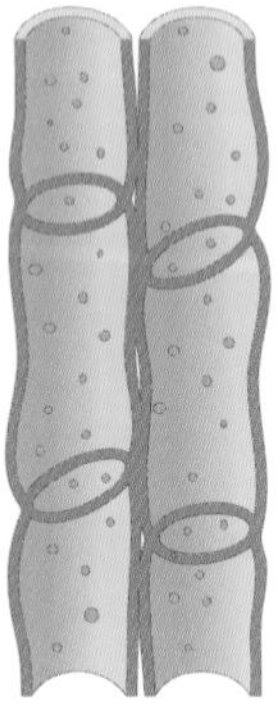

.................................................................. ..................................................................

25. State the name of the different tissues in the following diagrams.

**Monocotyledon stem**

..........................................................................................................

..........................................................................................................

..........................................................................................................

..........................................................................................................

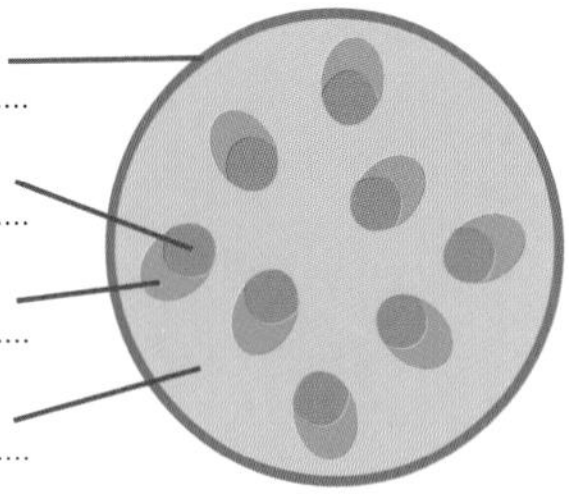

**Dicotyledon stem**

..........................................................................................................

..........................................................................................................

..........................................................................................................

..........................................................................................................

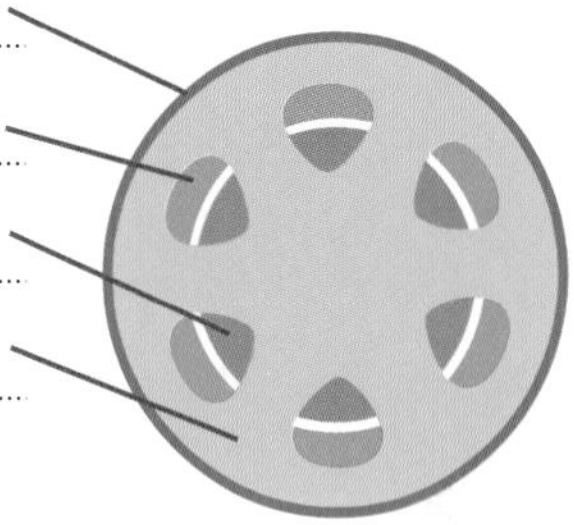

**Dicotyledon root**

..........................................................................................................

..........................................................................................................

..........................................................................................................

..........................................................................................................

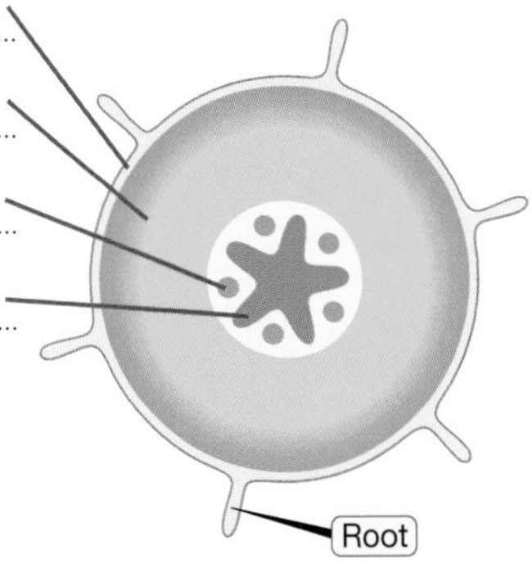

Root

**Leaf**

..........................................................................................................

..........................................................................................................

..........................................................................................................

..........................................................................................................

..........................................................................................................

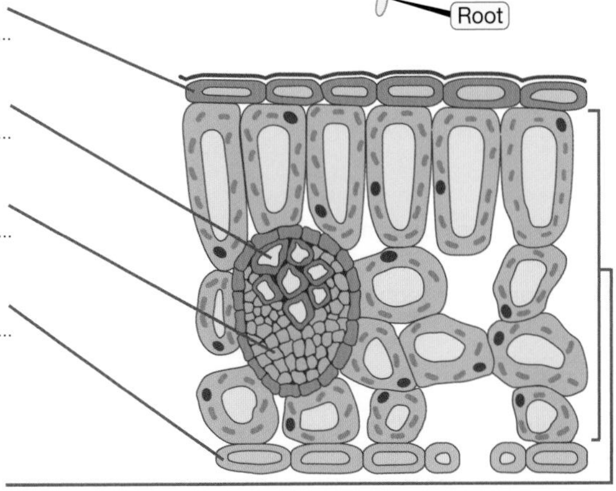

26. Name parts A, B and C of phloem tissue as shown in the diagram.

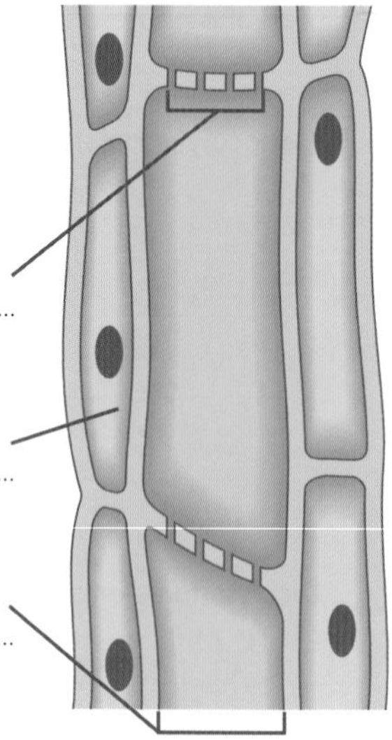

A ..............................................................................................................................

B ..............................................................................................................................

C ..............................................................................................................................

What is the function of part B and C?

B ..............................................................................................................................

..............................................................................................................................

C ..............................................................................................................................

..............................................................................................................................

27. A cross-section of a stem is shown below.

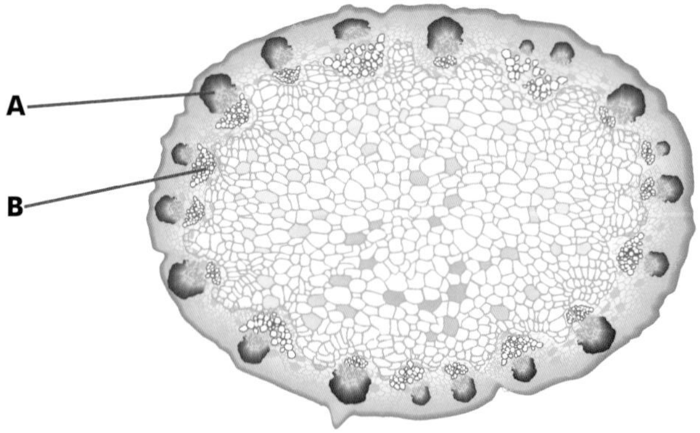

A

B

What type of stem is it? ..............................................................................................................................

Name one feature shown in the photograph that allows you to identify the section as a stem and not a root ..............................................................................................................................

Name the two vascular tissues, A and B, found in a vascular bundle.

A .............................................................. B ..............................................................

Draw a labelled diagram to show a longitudinal section of tissue A.

Include the following labels in your diagram: sieve tube, sieve plate and companion cell.

28. Give one function of each of the following:

    Dermal tissue ..........

    Ground tissue ..........

29. In which of the vascular tissues does water transport occur? ..........

    State one way in which this tissue is adapted for water transport ..........

    ..........

30. What kind of stem is shown in the diagram?

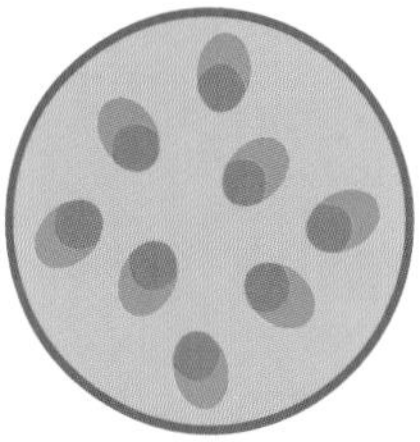

    ..........

    Give a reason for your answer ..........

    ..........

31. What part of the plant is shown in the diagram?

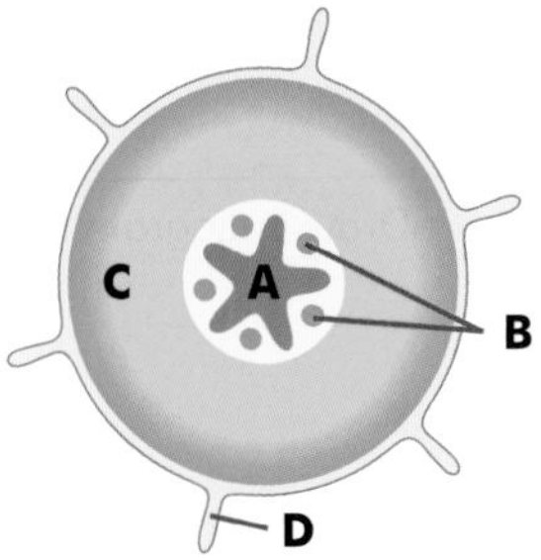

    ..........

    Give a reason for your answer ..........

    ..........



    Name the tissue type

    A ..........

    B ..........

    C ..........

    What is the function of part D? ..........

32. What kind of leaf is shown in the diagram?

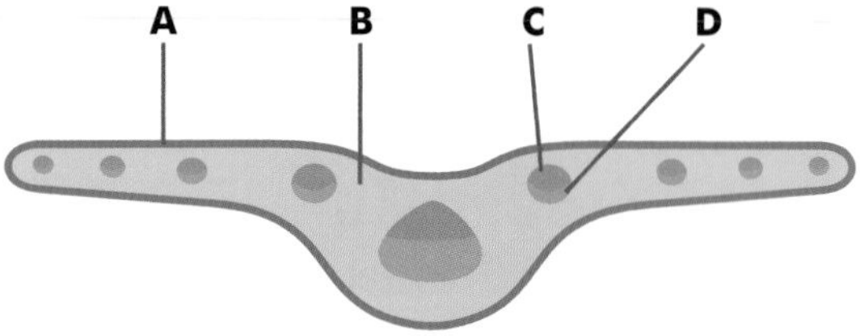

A B C D

Give a reason for your answer......................................

.......................................................................................

.......................................................................................

Name the substance carried in:

C.......................................................................................

D.......................................................................................

Name the tissues:

A.......................................................................................

B.......................................................................................

33. Name the different regions in the root tip of a plant.

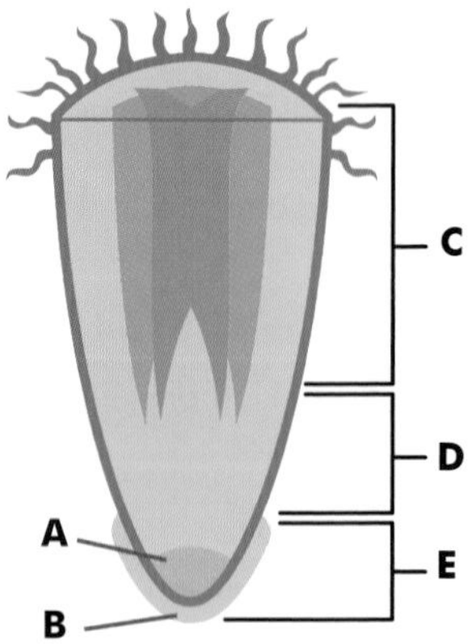

A.......................................................................................

B.......................................................................................

C.......................................................................................

D.......................................................................................

E.......................................................................................

C D A E B

What is the function of the root cap? ..............................................

.......................................................................................

Draw a diagram of the shoot tip.

# Unit 3

# Circulatory Systems

1. Name the two types of circulatory system and give one example of an animal you would find each system in.

   System ..........

   Example ..........

   System ..........

   Example ..........

2. Give three functions of blood.

   a) ..........

   b) ..........

   c) ..........

3. List the parts (constituents) of blood and give one function for each constituent.

| **Constituent** | **Function** |
|---|---|
| | |
| | |
| | |
| | |

4. What is blood plasma and state its function? ..........

   ..........

5. Name two types of cell found in the blood and give a function for each of them.

   Name .......... Function ..........

   Name .......... Function ..........

6. The ABO blood group system has four blood groups. What are the four groups?

   ..........

   ..........

Suggest a reason why it is important to know a person's blood group ..........

..........

7. Give one function for each of the following:

   Artery ..........

   Vein ..........

   Capillary ..........

8. List two differences between arteries and veins.

| **Artery** | **Vein** |
|---|---|
| a) .......... .......... | a) .......... .......... |
| b) .......... .......... | b) .......... .......... |

9. Write the correct name under each diagram.

.......... .......... ..........

10. a) Name the two blood vessels shown in the diagram below and (b) name the layers in the wall of each vessel.

A: .......... B: ..........



..........

..........

..........

11. Answer the following questions in relation to blood vessels in the human body.

    Valves are present in veins. What is their function? ..........

    ..........

    Why are valves not needed in arteries? ..........

    ..........

Which has the bigger lumen (cavity), an artery or a vein? ..........

The wall of capillaries is only one cell thick. How is this related to their function? ..........

..........

How does a portal vein differ from other veins? ..........

..........

12. What is the function of the heart? ..........

13. Name the parts of the heart.

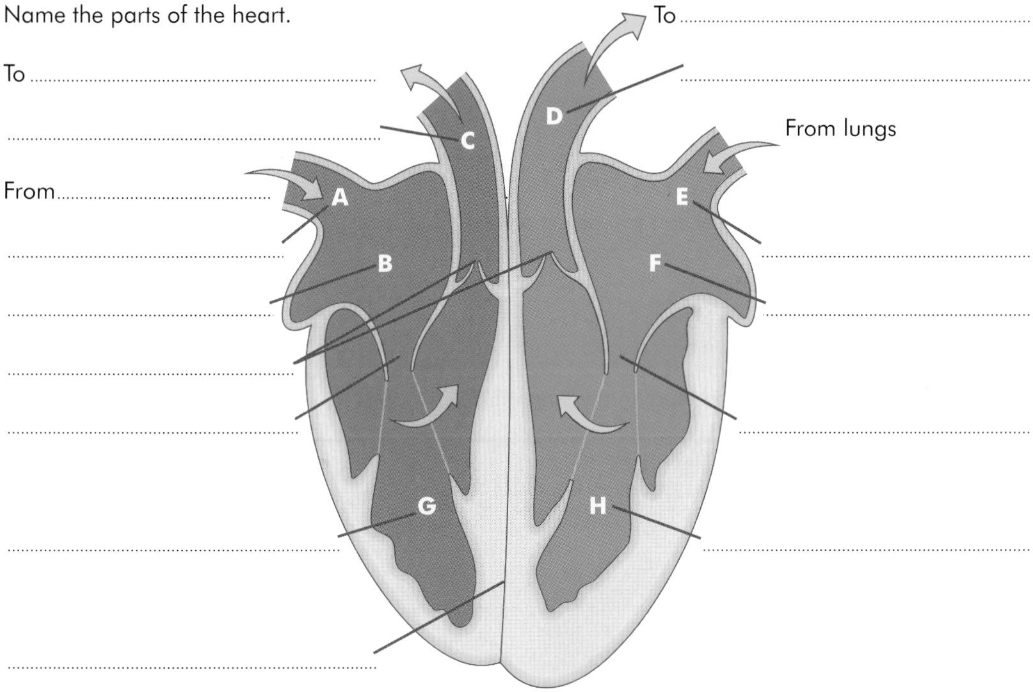

To ..........

..........

From ..........

..........

..........

..........

..........

..........

..........

To ..........

..........

From lungs

..........

..........

..........

..........

A B C D E F G H

Put the letters in the diagram above in the order blood flows through the heart.

..........

Name blood vessel **A** ..........

Is the blood in **A** oxygenated or deoxygenated? ..........

Give one reason why the wall of chamber **H** is thicker than the wall of chamber **G** ..........

..........

..........

What is the role of the bicuspid valve? ..........

..........

14. Name the blood vessel that carries blood to the heart muscle.

.......................................................................................................................

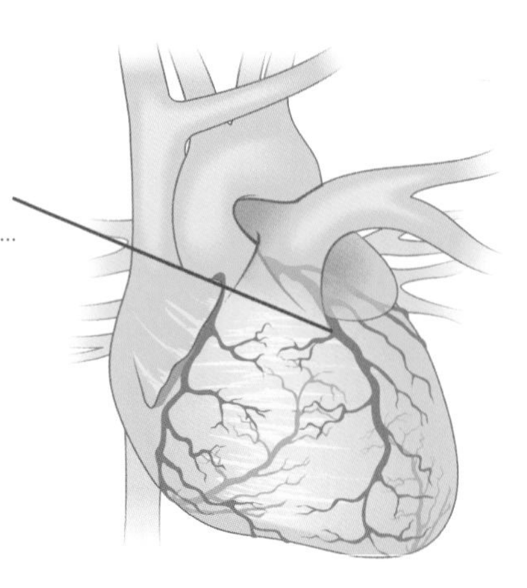

15. What is the normal heartbeat in an adult at rest? .......................................................................................................................

16. What is a pulse? .......................................................................................................................

17. In which type of blood vessel can you detect a pulse? .......................................................................................................................

18. What causes blood pressure? .......................................................................................................................

.......................................................................................................................

19. What is the typical blood pressure in a young adult? .......................................................................................................................

20. Humans have a double circulatory system.
Write in the name of the two systems in the diagram.

A .......................................................................................................................

B .......................................................................................................................

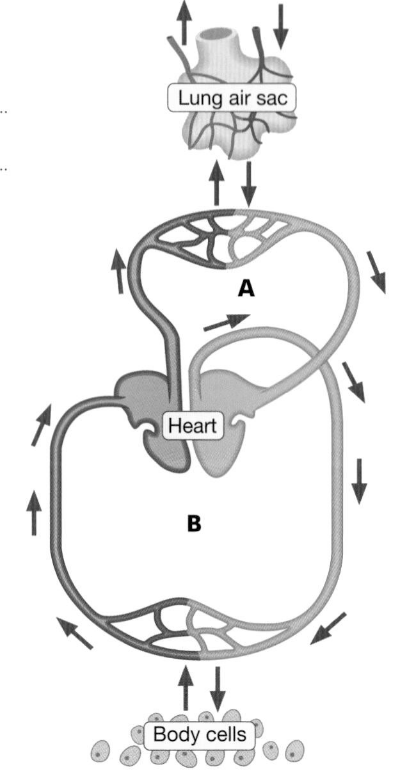

21. Name the main blood vessels shown in the diagram.

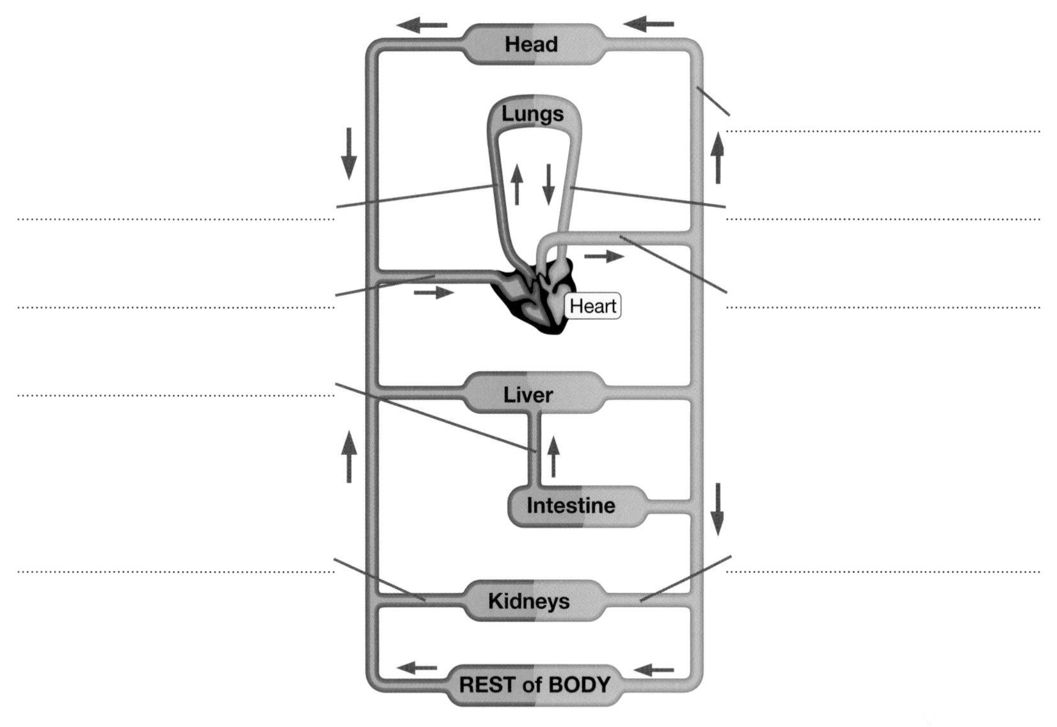


22. Name the chamber of the heart that receives blood back from the lungs.

.............................................................................................................

23. Name the blood vessels that bring this blood back from the lungs.

.............................................................................................................

24. Name the liquid part of blood .............................................................................

Name two parts of this liquid a)........................................ b)........................................

25. Complete the following table in relation to blood cells:

| Cell type | One function |
|---|---|
| Red blood cell | |
| White blood cell | |
| Platelet | |

26. Name the blood vessel referred to in each of the following:

a) The vein that connects the heart to the lungs ...............................................

b) The artery connected to the kidneys ...............................................

c) The vein that joins the intestine to the liver ..........

d) The artery that brings blood to the liver ..........

e) The vein that brings blood away from the liver ..........

27. Answer the following questions in relation to the human heart.

a) Give the precise location of the heart in the human body ..........

..........

b) What structure(s) protect(s) the heart? ..........

..........

c) Name the upper chambers of the heart ..........

..........

d) Name the valve between the upper and lower chambers on the left-hand side of the heart ..........

..........

28. Give **two** factors which cause an increase in heart rate.

a) .......... b) ..........

29. Name the blood vessels that bring oxygen to the heart muscle ..........

30. Explain why the walls of the lower chambers of the heart are thicker than those of the upper chambers.

..........

31. Use your knowledge of blood vessels and the information in diagrams below to fill in the table.

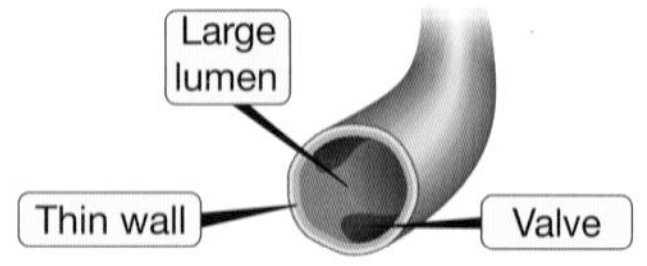


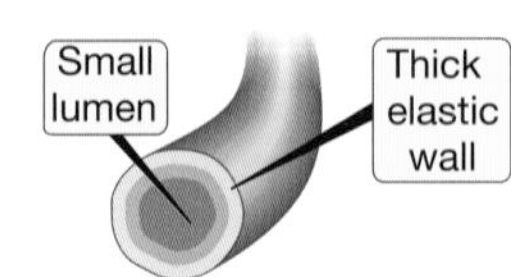


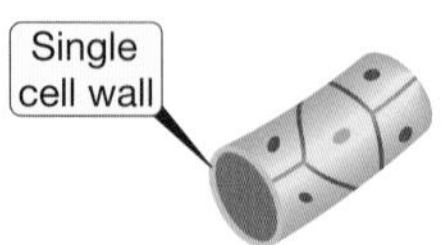


| | A | B | C |
|---|---|---|---|
| Name of vessel | | | |
| Wall | | | |
| Direction of blood flow | | | |
| Are valves present? | | | |
| Size of lumen | | | |

# Unit 3

# Lymphatic System

1. Give two functions of the lymphatic system.

   a) ..........

   b) ..........

2. Name the parts labelled A to G of the lymphatic system.

   A ..........

   B ..........

   C ..........

   D ..........

   E ..........

   F ..........

   G ..........

3. Name the liquid that surrounds the body cells ..........

   Where does this liquid come from? ..........

   What happens to the liquid that is not returned to the circulatory system? ..........

   ..........

4. What is the function of the lymph node? ..........

   ..........

5. Name one place in the human body where you find a lot of lymph nodes ..........

6. Name the two ducts where the lymph is returned to the blood. Also name the vein into which the lymph drains.

   ..........

   ..........

   ..........

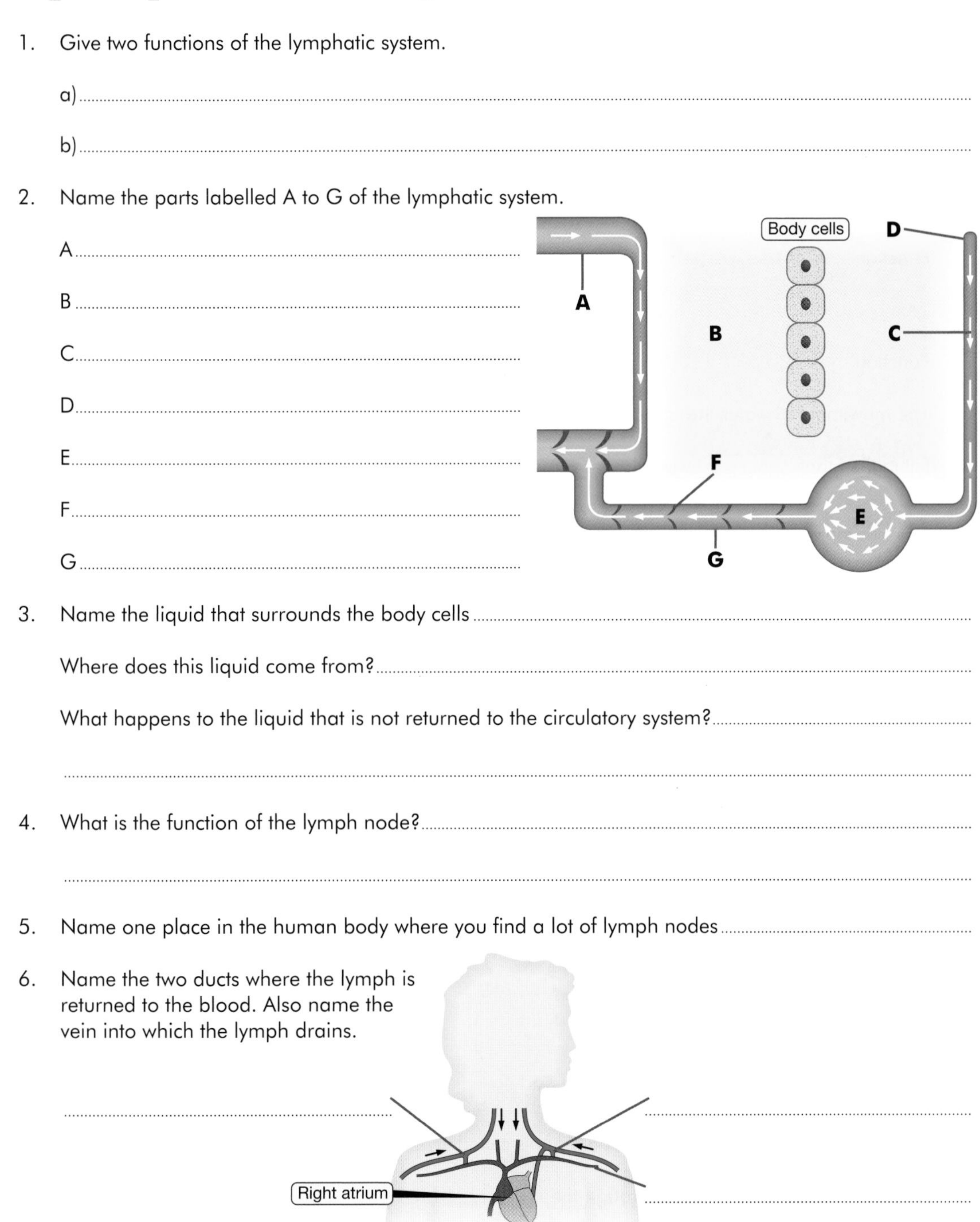

# Unit 3

## Transport in Plants

1. Why do plants need a transport system? ................................................

................................................

2. Name the two types of transporting tissue in plants and give one function for each type.

   Type ................................................

   Function ................................................

   Type ................................................

   Function ................................................

3. The movement of water through a plant is called ................................................

4. Fill in the blank spaces showing the movement of water through a plant.

   Soil water → ................................ into root ................................ →

   ................................ → into ................................ → up stem

   ................................ → leaf ................................ → leaf cells →

   ................................ → out through ................................

5. Water enters the plant through the root hairs by ................................................

6. Describe two ways water is helped to move up the stem.

   a) ................................................

   ................................................

   ................................................

   b) ................................................

   ................................................

   ................................................

7. The loss of water from the surface of a plant is called ................................................

8. Water is released from a plant through the ................................................ in the leaf.

9. State two features of a leaf that help to control water loss.

   a) ...................................................................................................

   ...................................................................................................

   b) ...................................................................................................

   ...................................................................................................

10. Name the cells that control the opening and closing of the pores in the leaf ..............................

11. Minerals are transported from the roots to all parts of the plant in ..............................

12. State two ways plant cells get carbon dioxide.

   a) ...................................................................................................

   ...................................................................................................

   b) ...................................................................................................

   ...................................................................................................

13. State what happens to the oxygen made during photosynthesis ..............................

   ...................................................................................................

14. Starch made during photosynthesis is carried to other parts of the plant in the ..............................

   ...................................................................................................

15. a) Mark in the level of water in the glass tube after a few days.

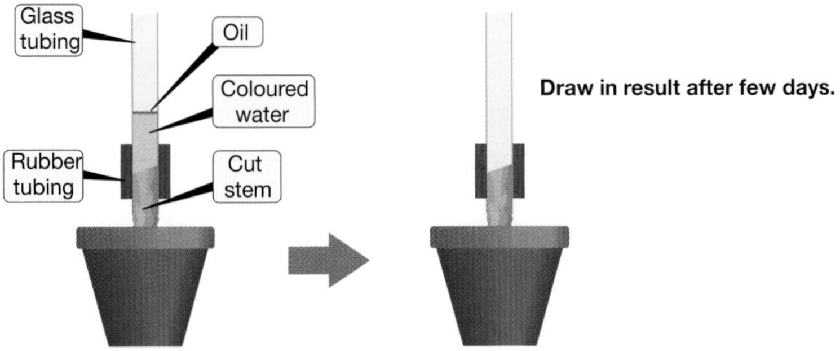

Glass tubing

Oil

Coloured water

Rubber tubing

Cut stem

**Draw in result after few days.**

   b) What is this process called? ..............................

16. What is the function of transpiration? ..............................

   ...................................................................................................

   ...................................................................................................

17. List two factors that increase the rate of transpiration.

    a) ..............................

    b) ..............................

18. What will form on the inside of the plastic bag?

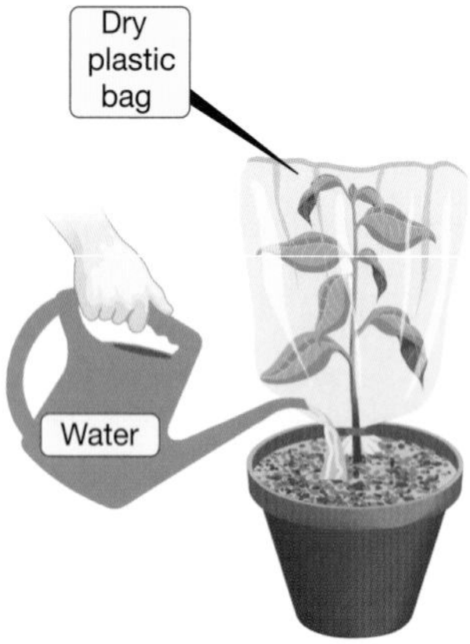

Dry plastic bag

Water

Where does the liquid come from? ..............................

This experiment is used to show which plant process? ..............................

19. Give one example of the following plant modifications:

    Root modifications ..............................

    Stem modifications ..............................

    Leaf modifications ..............................

20. Name the tissue that transports water from the root to the leaves ..............................

    Mention one way the tissue you have named is adapted for the transport of water ..............................

    ..............................

    ..............................

# Unit 3

# Nutrition and Digestion

1. What is an autotroph? ..............................

2. Give an example of autotrophic nutrition ..............................

3. Distinguish between autotrophic and heterotrophic nutrition ..............................

   ..............................

   ..............................

4. List the stages in human nutrition ..............................

   ..............................

5. Explain the following terms:

   Ingestion ..............................

   Digestion ..............................

   Absorption ..............................

   Egestion ..............................

6. Name the parts of the alimentary canal (digestive system)

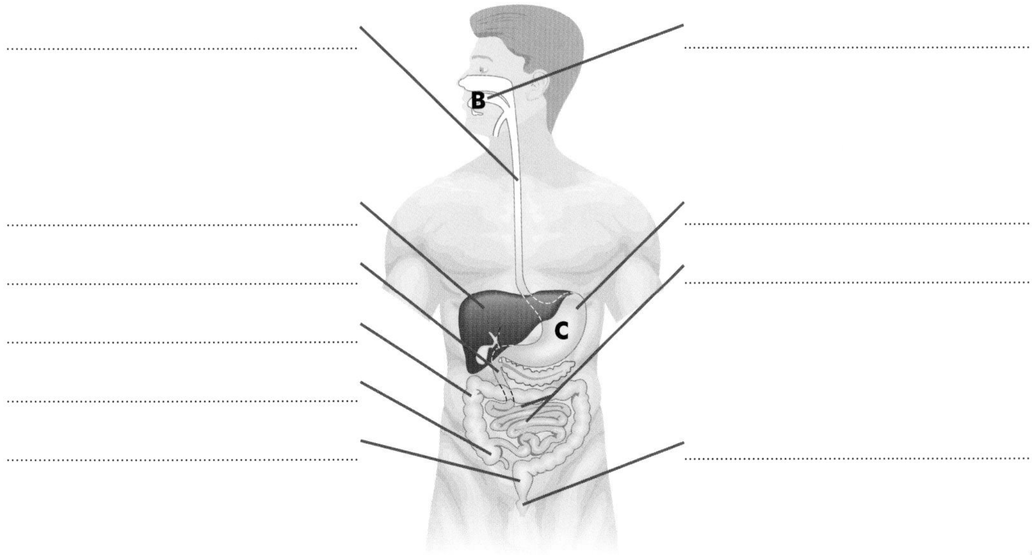

For each of the parts labelled B and C in the diagram of the alimentary canal, state whether the contents are acidic, neutral or alkaline.

B.......................................................................................... C..........................................................................................

7. Digestion is carried out with the help of chemicals called ..........................................................................

8. What is the function of peristalsis in the digestive system? ..........................................................................

..........................................................................................................................................................

9. Where do the products of digestion enter the blood? ..........................................................................

10. Amylase is an enzyme that is found in saliva. State the substrate and the product of this enzyme.

..........................................................................................................................................................

11. State two functions of symbiotic bacteria in the alimentary canal.

a) ..........................................................................................................................................................

b) ..........................................................................................................................................................

12. What is meant by egestion? ..........................................................................

Where does egestion take place? ..........................................................................

13. Name an enzyme that breaks down

Fats ..........................................................................................................................................................

Protein ..........................................................................................................................................................

14. Explain the term peristalsis ..........................................................................

..........................................................................................................................................................

15. What happens to food in the mouth? ..........................................................................

..........................................................................................................................................................

16. Name the enzyme present in the mouth ..........................................................................

What food does it breakdown? ..........................................................................

What is the end product? ..........................................................................

17. Name the juice in the stomach ..........................................................................

State two substances found in the juice and the function of each.

Substance 1 ..........................................................................................................................................................

Function ....................

Substance 2 ....................

Function ....................

Enzymes in the stomach need a .................... pH.

18. Name the two parts of the small intestine.

a) ....................

b) ....................

19. Name two digestive juices in the duodenum and state where each is made.

Juice 1 ....................

Made in ....................

Juice 2 ....................

Made in ....................

20. What are the functions of bile? ....................

....................

....................

Where is bile made? ....................

Where is bile stored? ....................

Where does bile act in the alimentary canal? ....................

21. Name an enzyme that acts on each of the following. State where it is made and the end products.

| | **Enzyme** | **Where it is made?** | **End products** |
|---|---|---|---|
| Starch | | | |
| Fat | | | |
| Protein | | | |

22. Where is the digested food absorbed? ....................

Give two adaptations that help absorption.

a) ....................

b) ....................

23. Name the structure in the diagram below and name the parts labelled C and D.

    Name ........................................

    C........................................

    D........................................

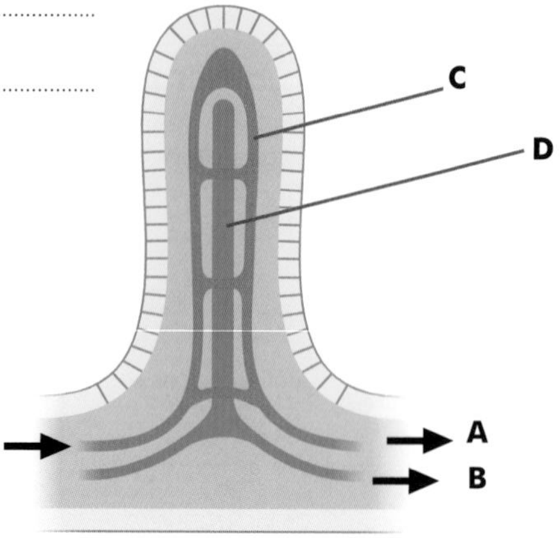

C
D
A
B

    To which location do vessels A and B deliver the absorbed digested material?

    A........................................ B........................................

24. Name the substance reabsorbed back into the body in the large intestine ........................................

25. What is the function of fibre in the diet?........................................

    ........................................

26. What is a balanced diet?........................................

    ........................................

27. Name the blood vessel that carries the digested food to the liver........................................

28. What are the following food constituents broken down into during digestion?

    Also state the vessel the end products are absorbed into in the small intestine.

| Constituent | End product(s) of digestion | Vessel absorbed into |
|---|---|---|
| Starch | | |
| Fat | | |
| Protein | | |

29. Name the two blood vessels that carry blood to the liver.

    a)........................................

    b)........................................

30. Give four functions of the liver.

    a) ........................................

    b) ........................................

    c) ........................................

    d) ........................................

31. In the liver, excess glucose is changed into ........................................

32. Name the waste product made by the liver from the breakdown of excess protein ........................................

33. Name a vitamin stored in the liver ........................................

34. How many teeth does an adult have? ........................................

35. Name each tooth type.

........................................ ........................................

36. Name the different types of teeth and state the function of each.

| Tooth type | Used for |
|---|---|
| | |
| | |
| | |

37. Name the types of teeth shown in the diagram below.

........................................

........................................

........................................

........................................

38. Write out the dental formula for an adult human ..........................................

..........................................

39. Fill in the blank spaces in the sentences below using the correct term from the following list:

**molar teeth, symbiotic bacteria, peristalsis, bile salts, lipase, stomach**

An organ for churning of food to chyme ..........................................

Waves of contractions passing along the gut ..........................................

Grind food into smaller pieces ..........................................

An enzyme that turns fats to fatty acids and glycerol ..........................................

Emulsify fats ..........................................

Produce vitamins ..........................................

40. Fill in the blank spaces.

The passage of the products of digestion from the intestine to the blood is called ...................................

Finger-like projections in the lining of the intestine called ..........................................

.......................................... increase the surface area for this passage. Amino acids from the digestion of .......................................... and monosaccharides from the digestion of .......................................... enter the blood in this process.

# Homeostasis

1. What is homeostasis? ..................................................

..................................................

2. Why is homeostasis necessary in animals? ..................................................

..................................................

a) Name two conditions that are controlled in homeostasis ..................................................

..................................................

..................................................

b) How is homeostasis achieved? ..................................................

..................................................

..................................................

3. Name three organs of the body that are involved in homeostasis and briefly explain what they do.

Organ .................................. What it controls ..................................

Organ .................................. What it controls ..................................

Organ .................................. What it controls ..................................

4. Fill in the missing words. If there was not a constant .................................. environment, our .................................. would not work properly. This would mean that nothing would operate properly and we would ...................................

5. In large organisms, the problem of size in relation to diffusion and gas exchange has been overcome by various means. Name two of them.

a) ..................................................

b) ..................................................

6. Why do multi-celled animals need a circulatory system? ..................................................

..................................................

..................................................

# Unit 3

# Gas Exchange in Flowering Plants

1. Exchange of gases in the leaf takes place through ..........

2. The underside of the leaf contains a large number of ..........

3. During the day photosynthesis takes place in the leaf cells. Carbon dioxide diffuses in through the .......... and oxygen diffuses out through the ...........

4. Water vapour also diffuses out through the .......... during transpiration.

5. a) Name the structure shown in the diagram below.......... In which part of the plant would you find the structure shown in the diagram? ..........

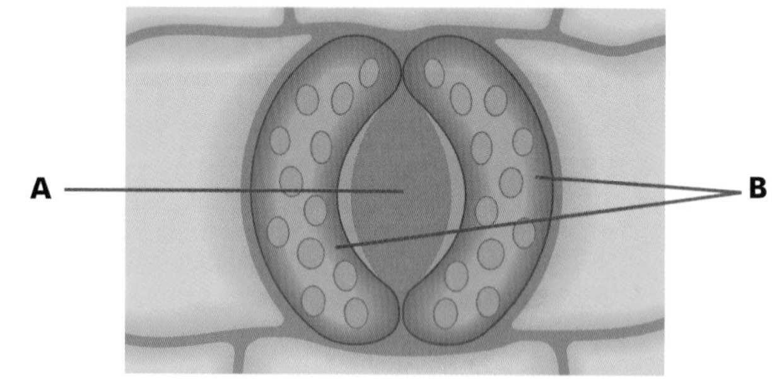


b) Name the parts A and B.

A.......... B..........

6. Name the structures found on the stem where gas exchange takes place.

..........

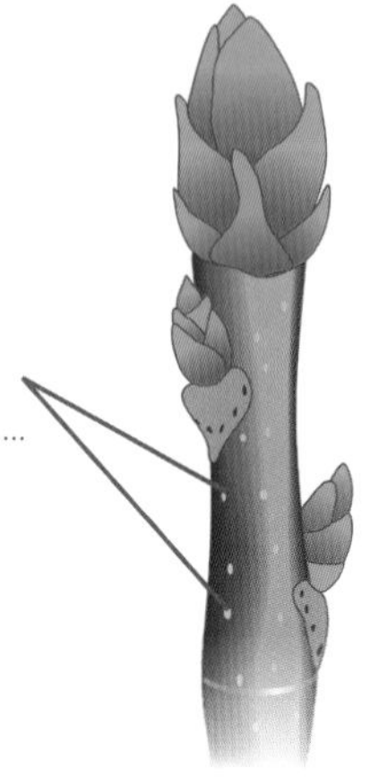

7. Name the parts of the leaf labelled A to G in the diagram below.

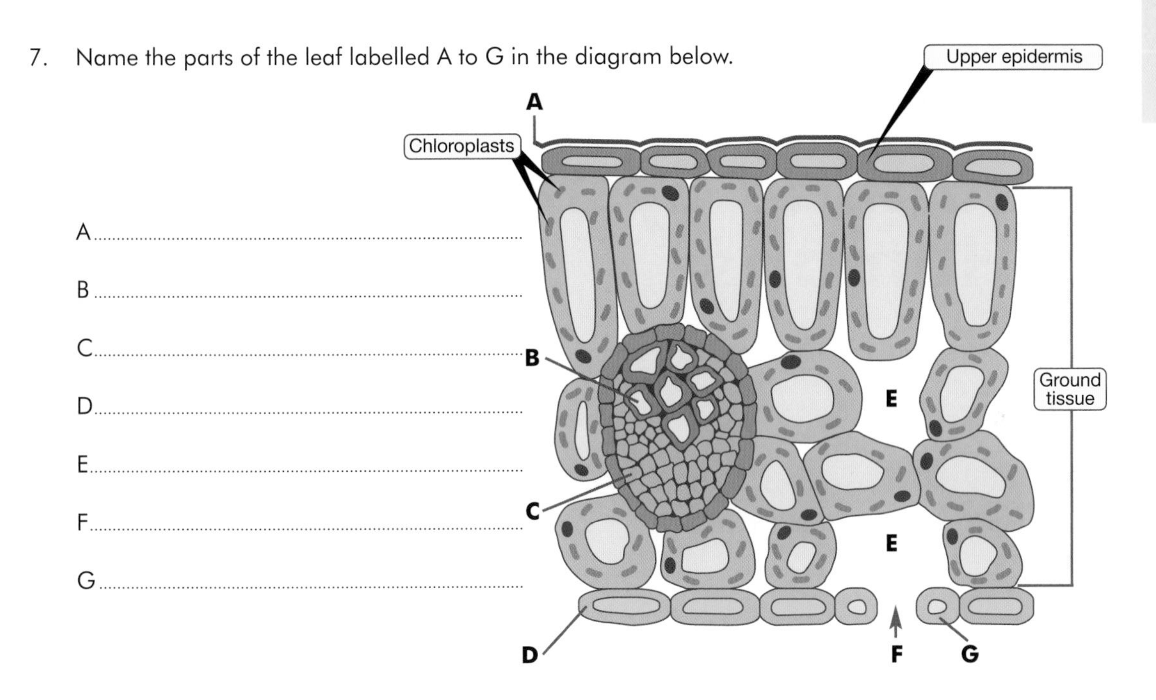


A.......................................................................

B.......................................................................

C.......................................................................

D.......................................................................

E.......................................................................

F.......................................................................

G.......................................................................

8. Each stoma has two ............................................................ on each side of the opening.

These cells ............................................................ and close stoma by changing their shape.

# Unit 3

# Breathing System in Humans

1. Give two reasons why animals need a breathing system.

   a) ..............................

   b) ..............................

2. List two features of a breathing system.

   a) ..............................

   b) ..............................

3. Name the parts of the breathing system.

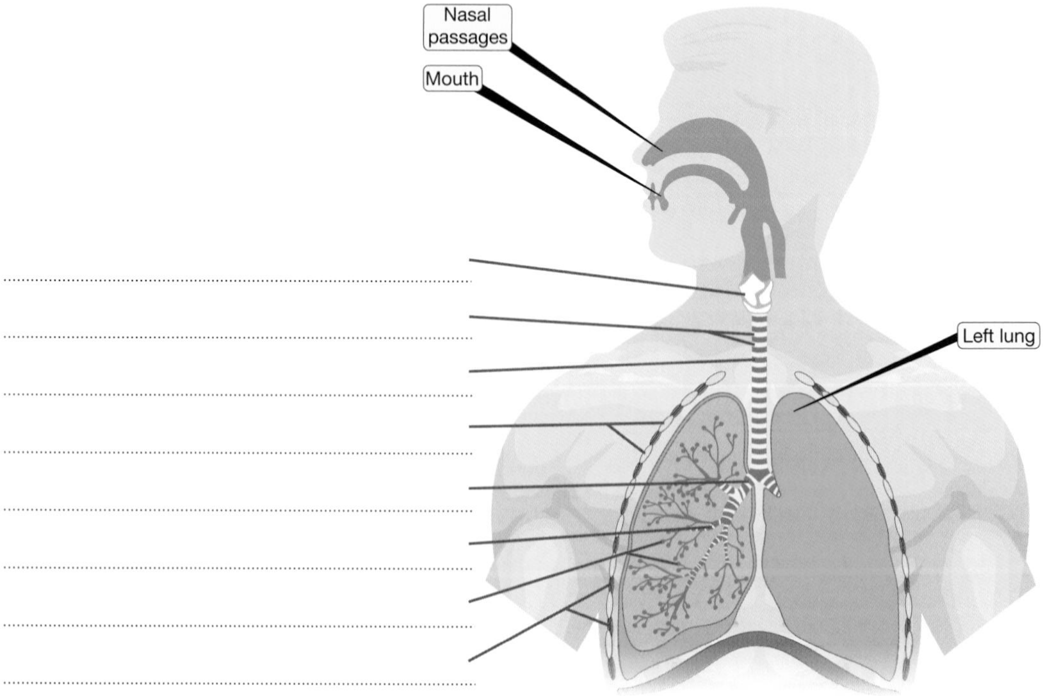

Nasal passages

Mouth

Left lung

..............................

..............................

..............................

..............................

..............................

..............................

..............................

..............................

4. Give two functions of the nose.

   a) ..............................

   b) ..............................

5. Where does the exchange of gases take place in the lungs? ..........

6. Name the structure shown .......... and fill in the blank spaces.

to .......... from ..........

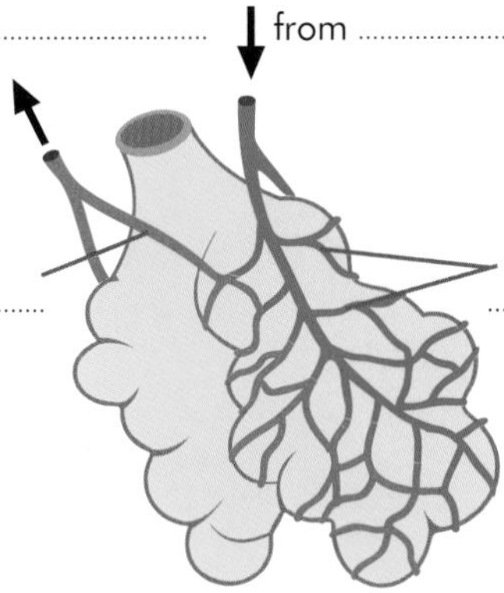

.......... ..........

7. What is the function of the epiglottis? ..........

8. What is the correct name for the voice box? ..........

9. What is the main function of the rings of cartilage in the windpipe? ..........

..........

10. Blood coming to an air sac is rich in ..........

11. Exhaled air is rich in ..........

12. Name the location of the intercostal muscles ..........

What is the function of the intercostal muscles? ..........

..........

13. Name the location of the pleural membranes ..........

..........

14. What is the function of the liquid between the pleural membranes? ..........

..........

15. Describe the process of breathing in (inhaling) ..........

..........

..........

..........

..........

16. Draw in the diagram the shape of the balloons when the rubber sheet is pulled down.

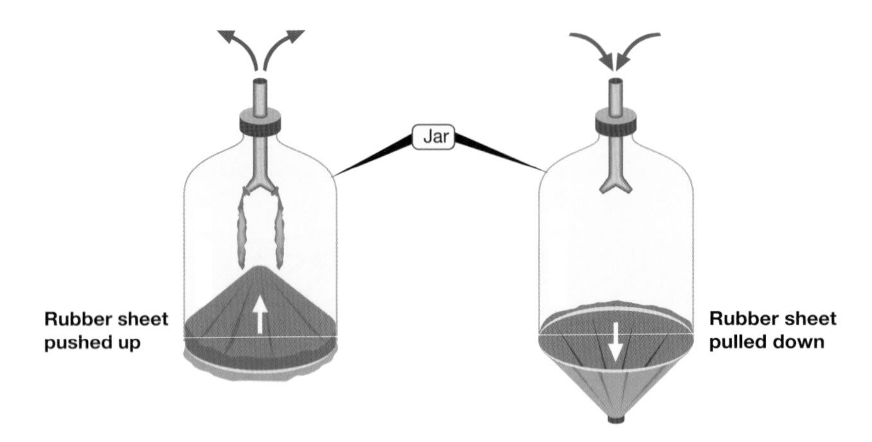


17. Oxygen is carried in the red blood cells by the red chemical ..........

18. Carbon dioxide is carried dissolved in the ..........

19. Give the cause, symptom, prevention and treatment for asthma.

    Cause ..........

    Symptom ..........

    Prevention ..........

    Treatment ..........

20. What effects does exercise have on breathing rate? ..........

    Give a reason for your answer ..........

    ..........

    ..........

21. What is the role of alveoli in the lungs? ..........

22. Name a breathing disorder ..........

# Excretion

1. Excretion is ..............................................................................................................................

2. Name two wastes made by the body.

   a) ..............................................................................................................................

   b) ..............................................................................................................................

3. Name the organs of excretion and state what is excreted from each.

| Organ | Waste removed |
|---|---|
| | |
| | |
| | |

4. What is the role of each of the excretory organs in homeostasis? ..............................................................................................................................

   ..............................................................................................................................

   ..............................................................................................................................

   ..............................................................................................................................

5. Name the parts of the skin shown in the diagram.

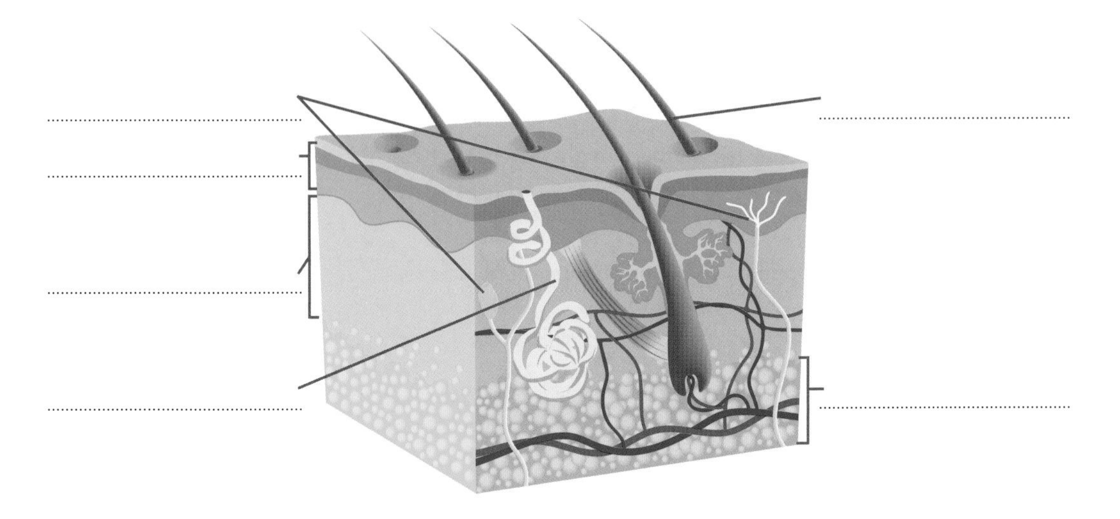

Give three functions of the skin.

a) ................................................................................................

b) ................................................................................................

c) ................................................................................................

Place **X** on the diagram to show where sweat reaches the skin surface.

Apart from water, name one other substance which is found in sweat ................................

6. Describe briefly one way the skin helps to retain heat in cold conditions ................................

................................................................................................

................................................................................................

7. Name two glands found in the skin and state what each gland produces.

Gland ................................................................................................

Substance made ................................................................................................

Gland ................................................................................................

Substance made ................................................................................................

8. Name two types of sensory receptors in the skin.

a) ................................................................................................

b) ................................................................................................

9. What is an endotherm? ................................................................................................

10. Give one example of an endotherm ................................................................................................

11. What is normal body temperature in humans? ................................................................

12. Give one reason why body temperature might increase ................................................

13. What is the main source of body heat in endotherms? ................................................

14. State two things that happen when body temperature increases.

a) ................................................................................................

................................................................................................

b) ................................................................................................

................................................................................................

# Unit 3

# Urinary System in Humans

1. Give two functions of the urinary system.

   a) ..........

   b) ..........

2. Where is urea made? ..........

3. Urea is made as a result of the breakdown of excess ..........

4. Name the parts of the urinary system shown in the diagram.

   ..........

   ..........

   ..........

   ..........

   ..........

   ..........

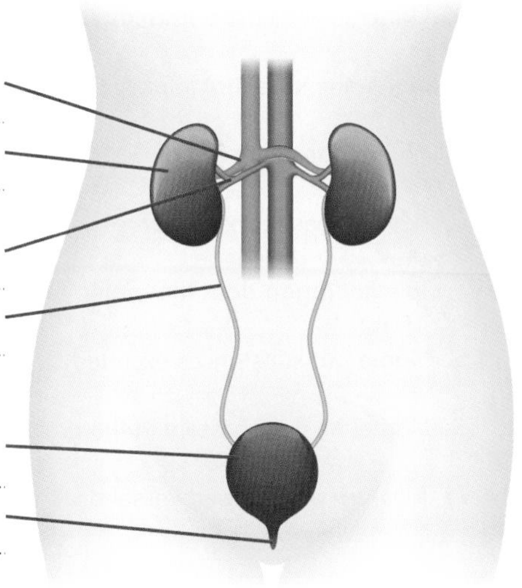

5. Name the parts of the kidney shown in the diagram.

   ..........

   ..........

   ..........

   ..........

   ..........

   ..........

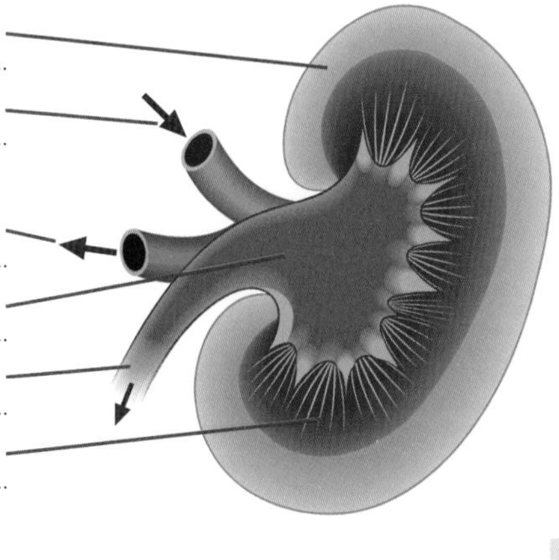

6. The making of urine takes place in stages. Name the stages and where they take place.

   Stage .................................................. location ..................................................

   Stage .................................................. location ..................................................

7. Name the artery that brings blood to the kidney..................................................

8. What happens during reabsorption? ..................................................

   ..................................................

   ..................................................

   ..................................................

9. Where is urine stored in the body? ..................................................

10. Urine flows to the outside of the body through the ..................................................

11. Name the tube that joins the kidney to the bladder ..................................................

12. Where does filtration of blood take place? ..................................................

13. Where does reabsorption of salt take place? ..................................................

14. To what organ does the ureter link the kidney? ..................................................

15. Name two substances excreted by the kidney ..................................................

16. Name the fluid present in the ureter ..................................................

17. To which main blood vessel does the renal artery link the kidney? ..................................................

    ..................................................

# Unit 3

# Responses in Flowering Plants

1. What is a stimulus? ..........

..........

..........

2. What is a tropism? ..........

..........

..........

3. Name the different types of tropisms, explain the response and give an example of each type.

| Tropism | Growth response | Example |
|---|---|---|
| | | |
| | | |
| | | |
| | | |
| | | |

4. Name the type of tropism shown in the diagram below ..........

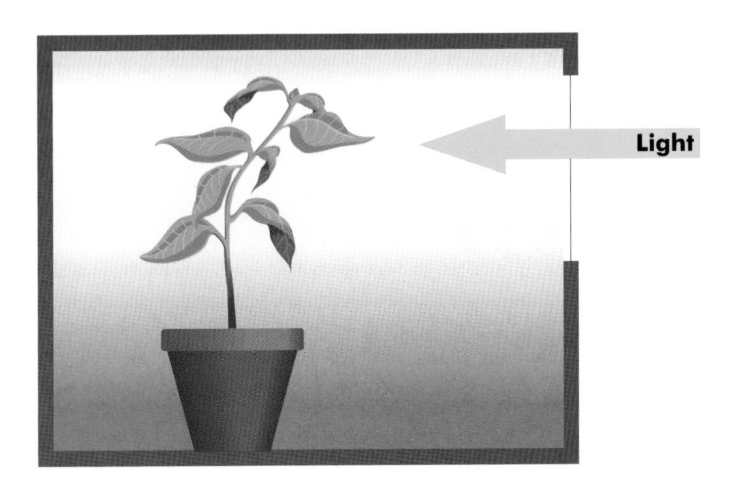

5. Name the type of tropism shown in the diagram below ..........

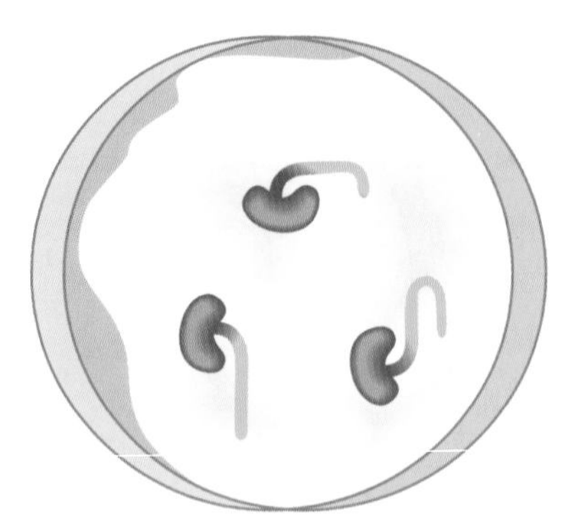

6. What is meant by a positive response to a stimulus? ..........

..........

..........

7. Phototropism is the response of a plant to ..........

Give one advantage of phototropism ..........

..........

..........

# Unit 3

# Growth Regulators

1. What is a plant growth regulator? ..........

..........

..........

2. Growth regulators are made in ..........

3. The meristematic regions are found at the tip of a ..........

   and the tip of a ..........

4. Why are growth regulators sometimes called plant hormones? ..........

..........

..........

..........

5. Name one type of growth promoter and one type of growth inhibitor.

   Growth promoter ..........

   Growth inhibitor ..........

6. State two uses for artificial growth regulators.

   a) ..........

   b) ..........

7. Name two structural features used by plants for protection.

   a) ..........

   b) ..........

8. Name one chemical method used by plants for protection ..........

..........

..........

..........

9. The diagram below is an example of which type of tropism? ........................................

   Which part of the plant shown in the diagram below has a positive response to the stimulus?

   ........................................

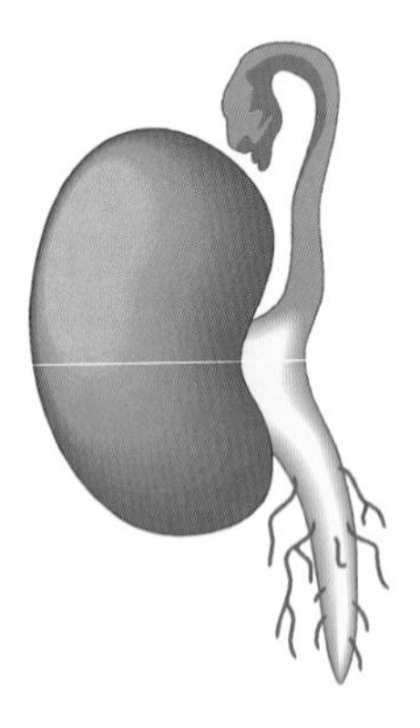

   Which part is stimulated by a low concentration of IAA? ........................................

   Which part is stimulated by a high concentration of IAA? ........................................

10. Give an example of a regulator that inhibits plant growth........................................

    ........................................

11. Give two uses of plant growth regulators in horticulture.

    a)........................................

    b)........................................

# Unit 3

# Responses in Humans/Nervous System

1. Name three systems responsible for responses in animals.

   a) ..............................

   b) ..............................

   c) ..............................

2. Name the two parts of the nervous system and what each consists of:

   Part 1 .............................. Consists of ..............................

   Part 2 .............................. Consists of ..............................

3. What do the letters CNS stand for? ..............................

4. What is a nerve cell called? ..............................

5. Name the three types of neuron and state the direction each carries nerve impulses.

   Name .............................. Direction of impulse ..............................

   Name .............................. Direction of impulse ..............................

   Name .............................. Direction of impulse ..............................

6. Name the parts labelled A to F in the diagram of a neuron.

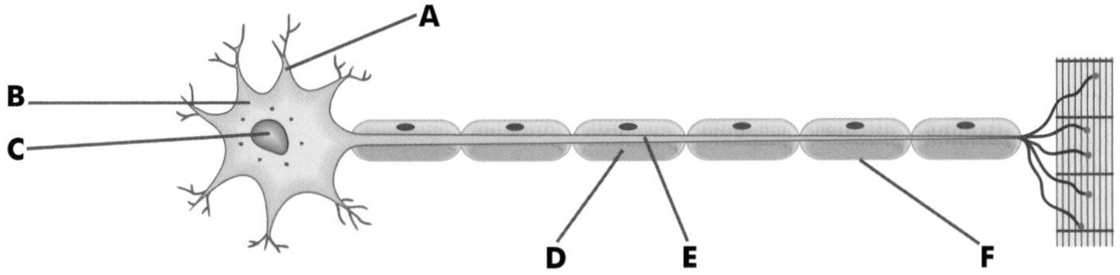

   A .............................. B ..............................

   C .............................. D ..............................

   E .............................. F ..............................

   What type of neuron is it? ..............................

7. Give one function for each of the following parts of a neuron.

   Cell body ..........

   Dendrites ..........

   Axon ..........

   Myelin sheath ..........

   Schwann cell ..........

8. The place where two nerve cells meet is called a ..........

   The impulse is carried across this gap by ..........

9. A nerve impulse is carried along a neuron by ..........

10. The diagram shows a ..........

    Name the parts shown in the diagram.

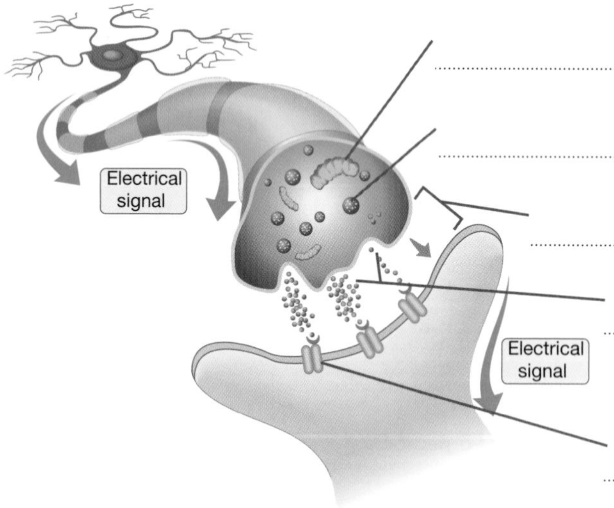

..........

..........

..........

..........

..........

11. Name the parts of the brain in the following diagrams.

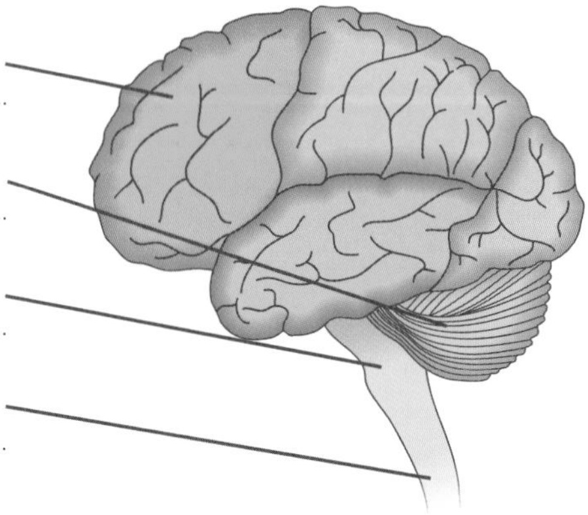

..........

..........

..........

..........

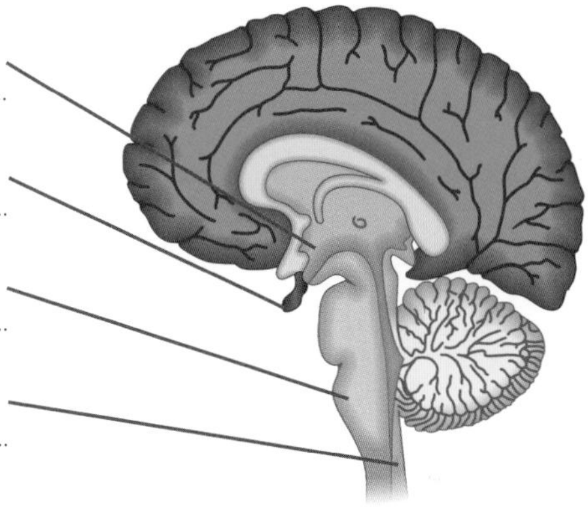

..............................................................................................................

..............................................................................................................

..............................................................................................................

..............................................................................................................

12. Give two functions of the brain.

a) ..............................................................................................................

b) ..............................................................................................................

13. Give one function for each of the following parts of the brain.

Cerebrum ..............................................................................................................

Hypothalamus ..............................................................................................................

Pituitary gland ..............................................................................................................

Cerebellum ..............................................................................................................

Medulla oblongata ..............................................................................................................

14. Give one cause, symptom, treatment and possible cure for Parkinson's disease.

Cause ..............................................................................................................

..............................................................................................................

..............................................................................................................

Symptom ..............................................................................................................

..............................................................................................................

..............................................................................................................

Treatment ..............................................................................................................

..............................................................................................................

Cure ..............................................................................................................

..............................................................................................................

15. Give two functions of the spinal cord.

    a) ..........

    b) ..........

    Name the parts of the spinal cord shown in the diagram.

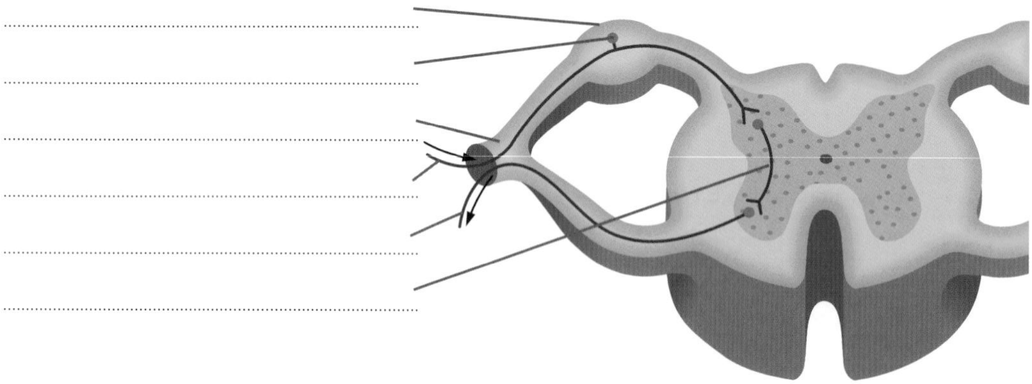

    ..........

    ..........

    ..........

    ..........

    ..........

    ..........

16. What is a reflex action? ..........

    ..........

17. Name the three nerves involved in a reflex action in the order the impulse is carried.

    a) ..........

    b) ..........

    c) ..........

18. Give an example of a reflex action ..........

    ..........

19. Name the parts and mark on the diagram the direction the impulse travels in a reflex action.

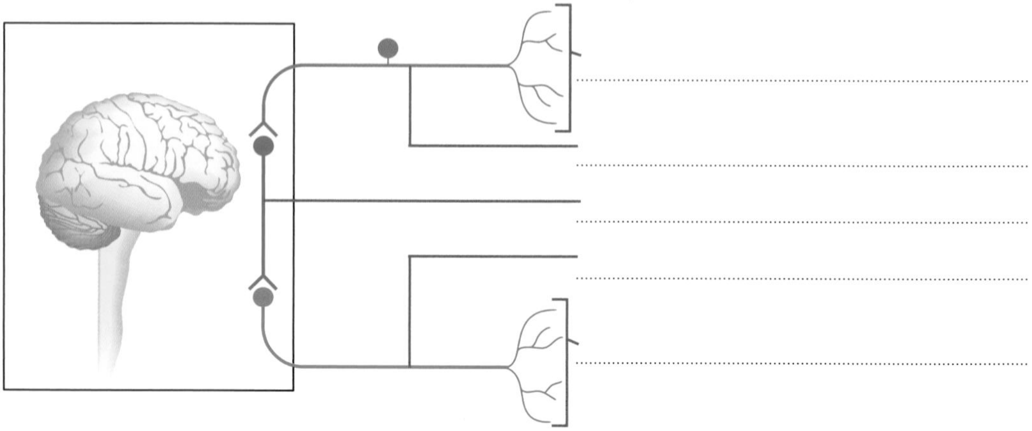

    ..........

    ..........

    ..........

    ..........

    ..........

20. Neurons produce neurotransmitter substances. What is their function? ..........

..........

..........

21. Describe the use the body makes of chemicals in the transmission of nerve impulses ..........

..........

..........

..........

22. What is the effect on the body if the spinal cord is broken? ..........

..........

23. What protects the spinal cord? ..........

24. Give an example of a reflex action in humans ..........

25. Why are reflex actions important in humans? ..........

# Unit 3

# Sense Organs

1. Name the sense organs and the sense each responds to.

| Sense organ | Sense |
|---|---|
| | |
| | |
| | |
| | |
| | |

2. Taste is detected by receptors in the tongue called ..........

3. Each taste bud responds to one of the four tastes. Name the four tastes.

   a) .......... b) ..........

   c) .......... d) ..........

4. Receptors in the skin together form the sense of ..........

5. Some parts of the skin have more receptors than others, for example ..........

6. Name the parts in the diagram.

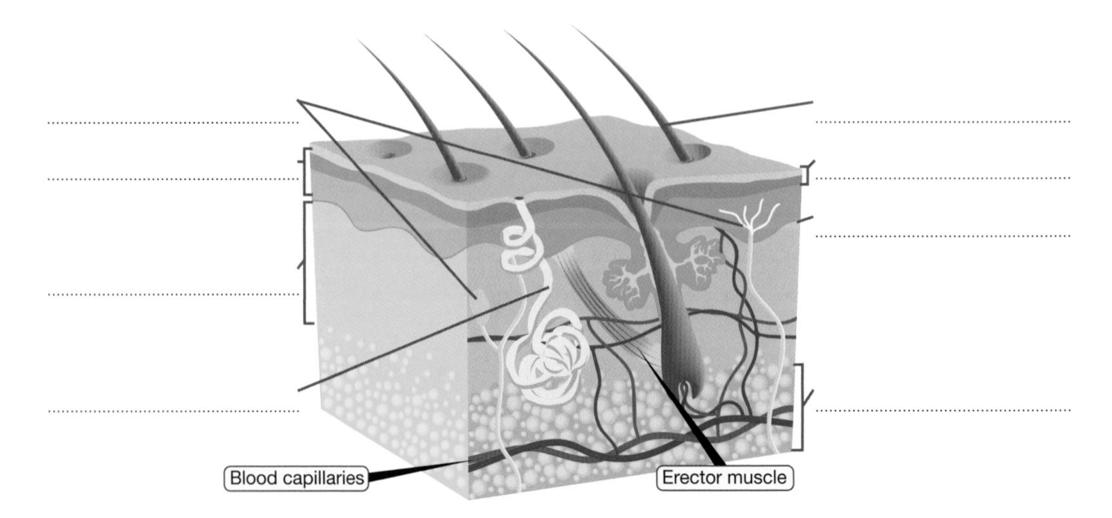

# Unit 3

## The Eye

1. Name two ways in which the eye is protected.

   a) ........................................................................

   b) ........................................................................

2. Name the parts of the eye shown in the diagram.

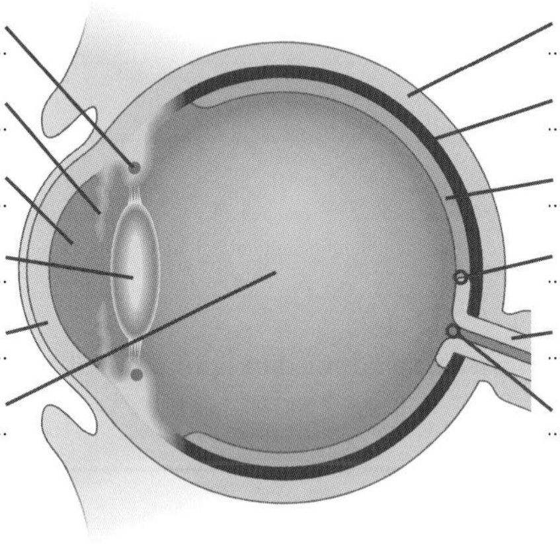

3. Give one function for each of the following:

   The sclera ........................................................................

   The cornea ........................................................................

   The iris ........................................................................

   The pupil ........................................................................

   The lens ........................................................................

   The ciliary muscle ........................................................................

   The retina ........................................................................

   The optic nerve ........................................................................

4. Name the two types of light sensors/receptors found in the retina.

   a) ........................................................................

   b) ........................................................................

5. Which sensor/receptor works in bright light and detects colour? ..................

6. Describe what happens when you want to focus on a distant object ..................

..................

..................

7. What is long sight? ..................

..................

How is it corrected? ..................

8. What is short sight? ..................

..................

How is it corrected? ..................

9. What is the function of the fovea? ..................

10. What is missing at the blind spot? ..................

11. Is the eye shown in the diagram adapted for dim light or bright light?

Explain your answer ..................

..................

12. Where in the eye is the retina located? ..................

13. Name the two types of light sensitive cells found in the retina.

a) ..................

b) ..................

14. Give one difference between the two types of light sensitive cells ..................

..................

15. The .................. nerve is attached to the eye.

What is its function? ..................

# Unit 3

# The Ear

1. Give two functions of the ear.

   a) ..............................

   b) ..............................

2. Fill in the parts of the ear shown in the diagram below.

..............................

..............................

..............................

..............................

..............................

..............................

..............................

..............................

..............................

..............................

3. Give one function for each of the following parts:

   Pinna (earlobe) ..............................

   ..............................

   Ear canal ..............................

   ..............................

   Eardrum ..............................

   ..............................

Ossicles ..............................................................................................................

..........................................................................................................................

Semi-circular canals .................................................................................................

..........................................................................................................................

Cochlea ..............................................................................................................

..........................................................................................................................

Auditory nerve ......................................................................................................

..........................................................................................................................

Eustachian tube ....................................................................................................

..........................................................................................................................

4. Sensory receptors in the ................................................ detect changes in the movement of the head.

5. The structure that separates the outer ear from the middle ear is called the ..........................................

..........................................................................................................................

6. Name parts A and B in the diagram and give the function of each.

A ..........................................................................................................................

Function ..............................................................................................................

B ..........................................................................................................................

Function ..............................................................................................................

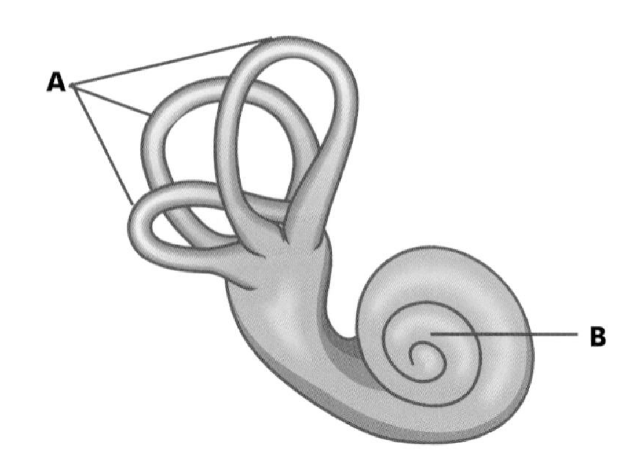

# Unit 3

# Endocrine System

1. The exocrine system consists of ..............................................................................................................

   ..............................................................................................................

2. Give an example of an exocrine gland ..............................................................................................................

3. The endocrine system consists of ..............................................................................................................

   ..............................................................................................................

4. What is a hormone? ..............................................................................................................

   ..............................................................................................................

   ..............................................................................................................

5. The hormones are carried around the body in ..............................................................................................................

6. What is a target organ? ..............................................................................................................

   ..............................................................................................................

   ..............................................................................................................

7. Name the organs of the endocrine system in the diagrams below.

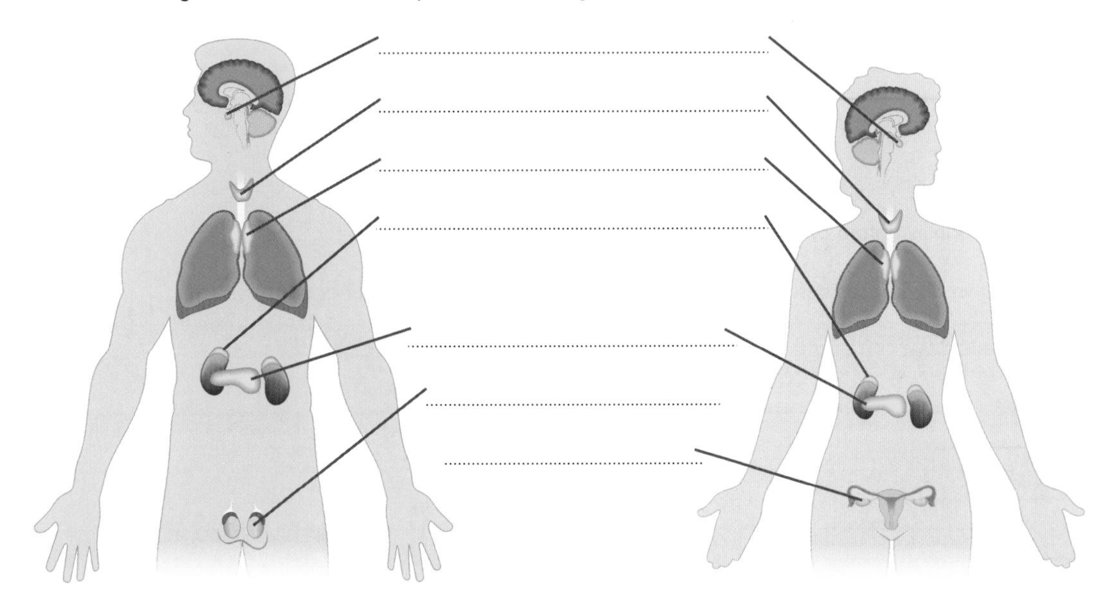

8. Fill in the blank spaces in the table below.

| Gland | Hormone | Function |
|---|---|---|
| | | |
| | | |
| | | |
| | | |
| | | |

9. Most hormones are made of ..........

10. The thyroid gland is in the neck and makes the hormone ..........

11. .......... is needed by the thyroid gland to make the hormone thyroxine.

12. Thyroxine controls ..........

13. The production of thyroxine by the thyroid gland is regulated by the ..........

14. A swelling of the thyroid gland is called.......... and it can be caused by

..........

..........

15. Give one symptom of the underproduction of thyroxine in

Children ..........

..........

Adults ..........

..........

16. Give one symptom of the overproduction of thyroxine ..........

17. List two differences between responses in the nervous system and endocrine system.

| Nervous response | Hormonal response |
|---|---|
| a) | a) |
| | |
| b) | b) |
| | |

# Unit 3

# Human Skeleton

1. List three functions of the human skeleton.

   a) ..........

   b) ..........

   c) ..........

2. Name the two parts of the skeleton.

   Name of part 1 .......... consists of ..........

   Name of part 2 .......... consists of ..........

3. What is the function of the skull? ..........

4. Name the parts of the skeleton shown in the diagram.

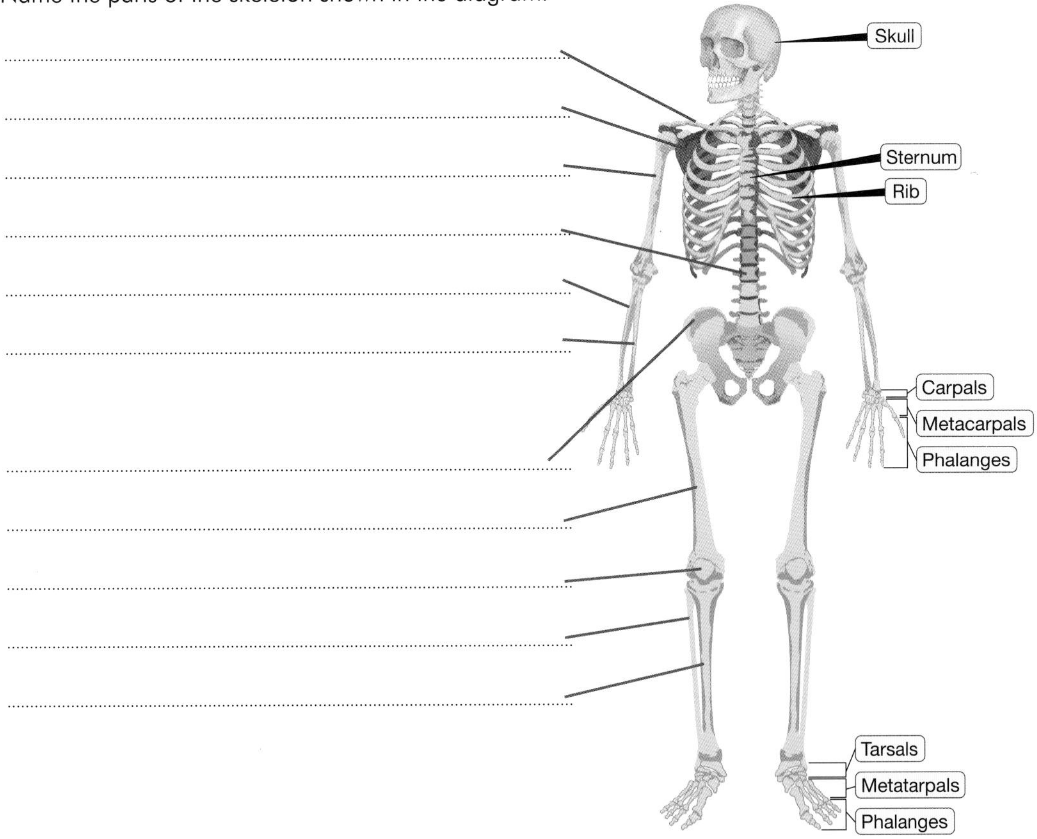

5. The small bones in the vertebral column (backbone) are called ..........

6. Name the regions of the vertebral column ..........

..........

7. The vertebral column protects the ..........

8. Name the two girdles in the body. a) .......... b) ..........

9. a) What is osteoporosis? ..........

..........

b) How can a person reduce the risk of developing osteoporosis? ..........

..........

10. Name the bones in the fore limb and hind limb.

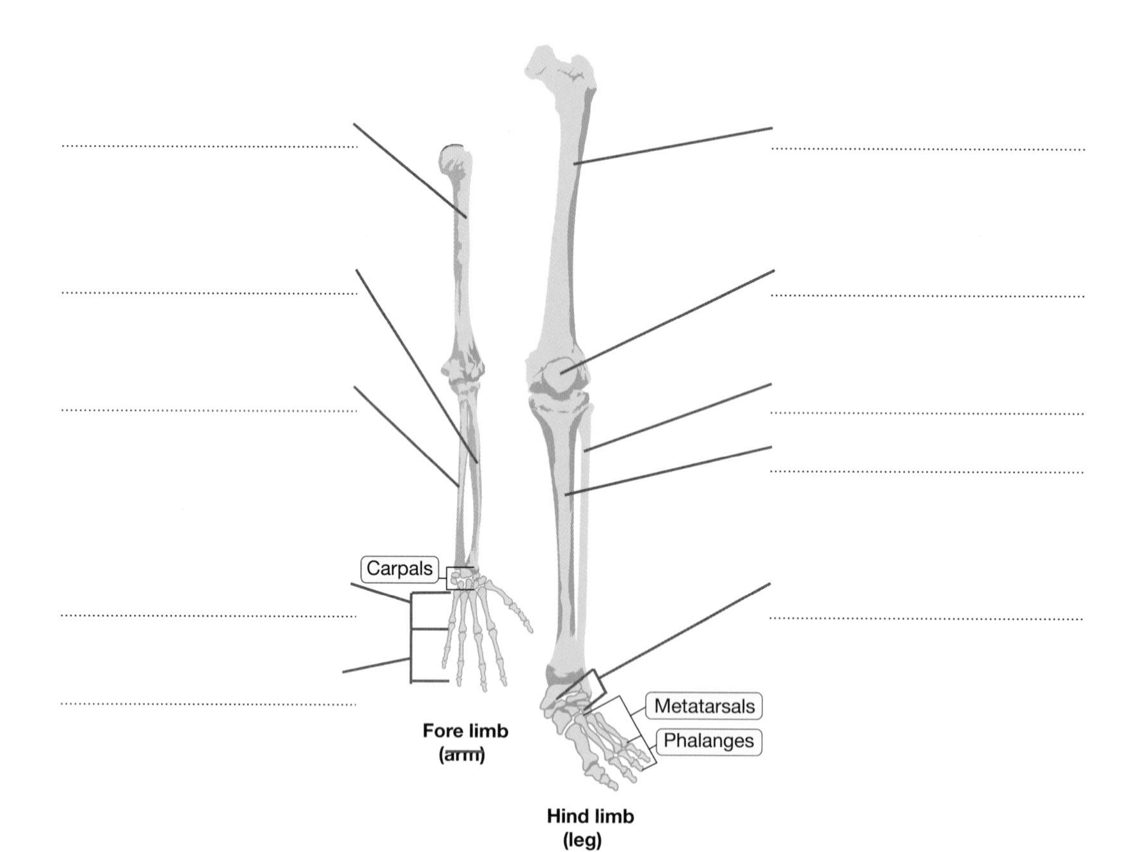

11. A long bone has a middle and two ends. The middle is called the ..........

 The cavity in the middle is called the ..........

 cavity and is filled with jelly-like ..........

 The ends of the bone are covered with ..........

 Give one function of this covering..........

12. Name the regions of the vertebral column and the number of vertebrae in each region.

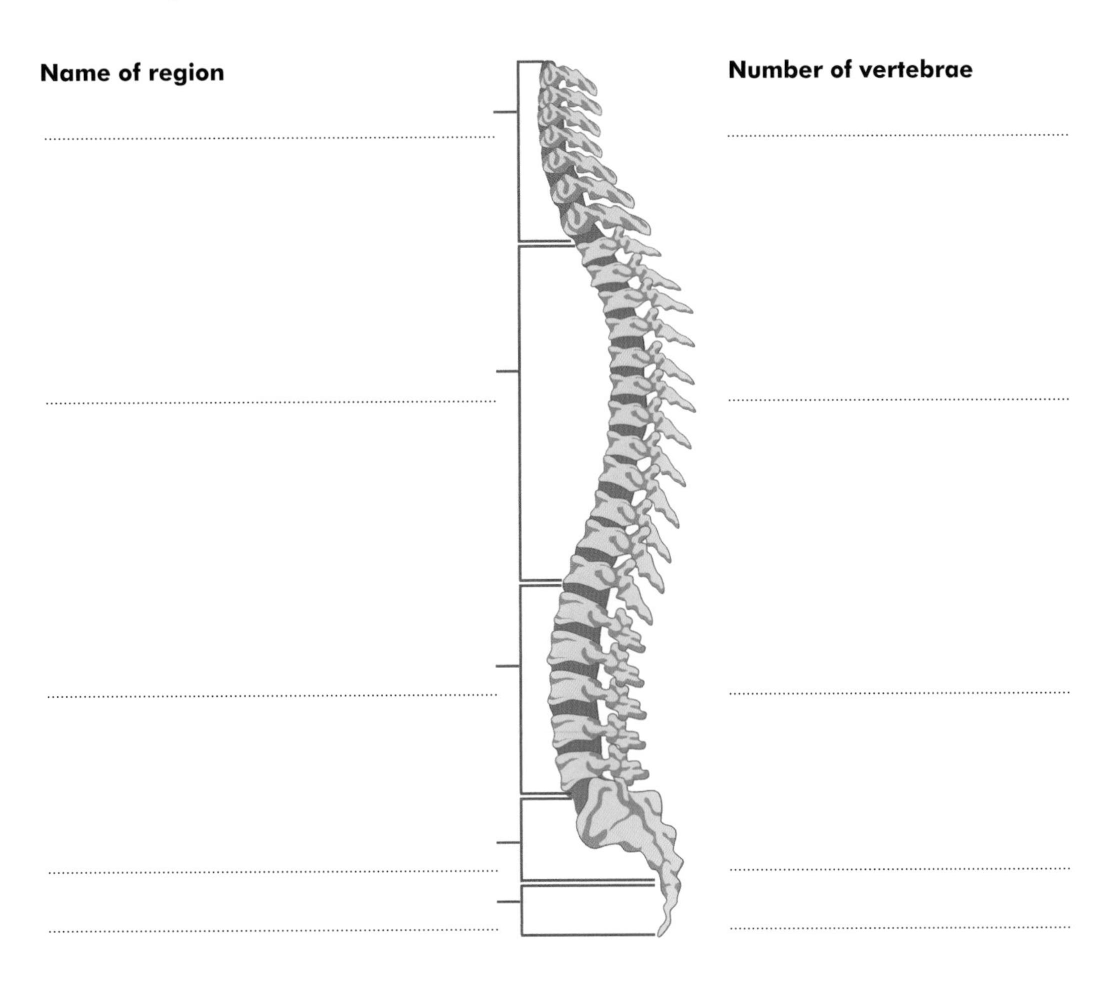


13. There are three types of joint in the body. Name them and give an example of each.

 a).......... example ..........

 b).......... example ..........

 c).......... example ..........

14. Name the parts of the joint in the diagram below.

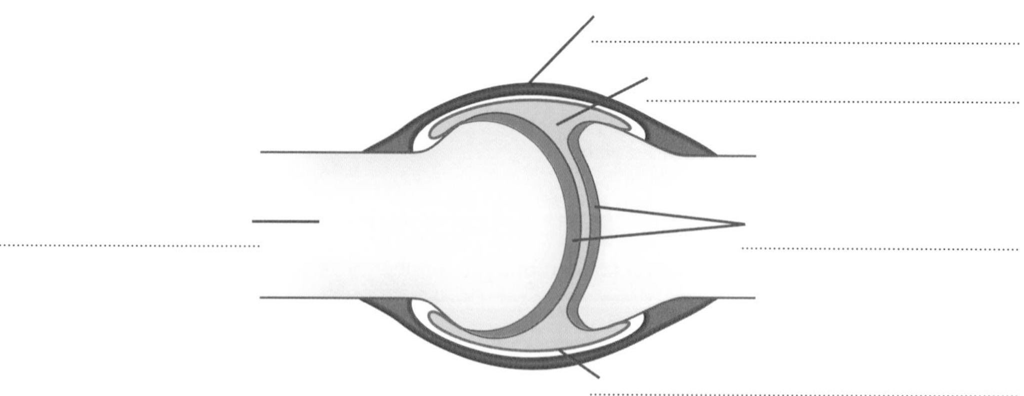

Give one function for each of the following parts:

Ligament ..........

Cartilage ..........

Synovial fluid ..........

Where in the body would you find a joint like the one shown in the diagram above? ..........

15. Give one cause, prevention and possible treatment for arthritis.

Cause ..........

..........

Prevention ..........

..........

Treatment ..........

..........

16. Name the vertebrae found in

The neck ..........

The small of the back ..........

17. Name the part of the central nervous system that runs through the vertebrae ..........

..........

18. Name the three main bones that form the human arm.

 a) ..........................................................................................

 b) ..........................................................................................

 c) ..........................................................................................

19. Name the parts labelled A to E of a long bone.

 A ......................................................

 B ......................................................

 C ......................................................

 D ......................................................

 E ......................................................

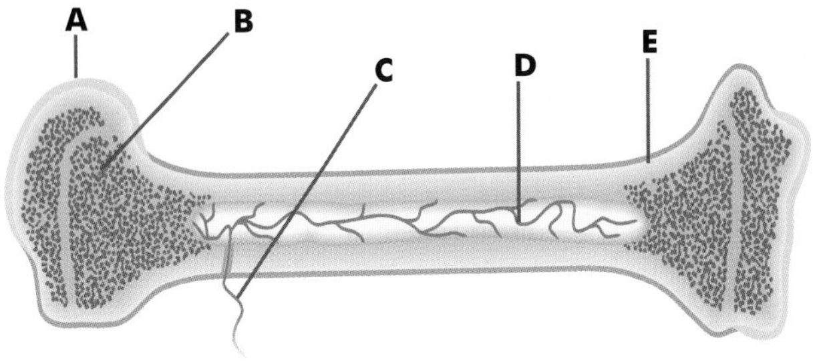

 What is the function of part A? ..........................................................................................

 What is found in part B? ..........................................................................................

 What is made in part B? ..........................................................................................

 Why is the bone not solid? ..........................................................................................

20. The human skeleton is divided into two parts known as:

 Part 1 ..........................................................................................

 made up of ...................................................... and ......................................................

 Part 2 ..........................................................................................

 made up of ...................................................... and ......................................................

21. Where are the discs in the human vertebral column? ..........................................................................................

22. What is the function of the discs? ..........................................................................................

23. Give a role for each of the following in the human body:

 a) Yellow bone marrow ..........................................................................................

 ..........................................................................................

 b) Red bone marrow ..........................................................................................

 ..........................................................................................

# Muscles

1. Name the three types of muscle.

   a) ..........

   b) ..........

   c) ..........

2. Muscles are joined to bones by ..........

3. When a muscle receives a nerve impulse it ..........

4. How does a muscle, joined to the skeleton, get back to its original shape after it contracts? ..........

   ..........

5. Skeletal muscles occur in opposing or antagonistic pairs. Explain the term antagonistic pair ..........

   ..........

   ..........

   ..........

   ..........

6. The diagram shows the muscles in the upper arm.

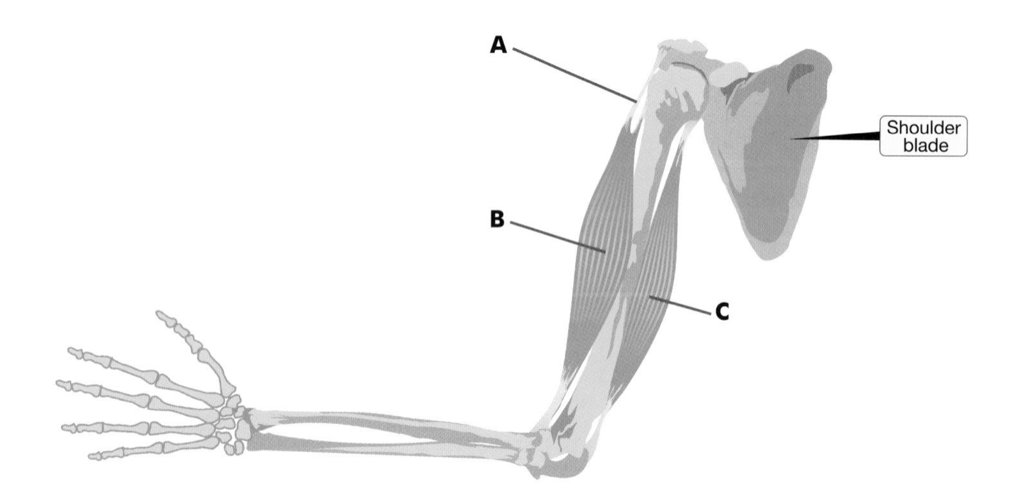

Name the parts labelled in the diagram.

A ......................................................................................................

B ......................................................................................................

C ......................................................................................................

Name the muscle that contracts to raise the arm ..........................................................

What is the function of part A? ....................................................................................

What type of muscle is C? ...........................................................................................

7. Smooth muscle is located in the ..............................................................................

   Example ....................................................................................................................

8. What is a voluntary muscle? ......................................................................................

   ..................................................................................................................................

9. Why is skeletal muscle sometimes called striped muscle? ...........................................

   ..................................................................................................................................

10. Cardiac muscle is found in the wall of the ................................................................

# Unit 3

## Viruses

1. What is a virus? ..........

2. Give one reason why viruses are not classified as living things ..........

..........

3. What is an obligate parasite? ..........

..........

4. The diagram shows a virus. Name the parts.

..........

..........

5. Viruses are classified according to shape. Name the three shapes ..........

   a) ..........

   b) ..........

   c) ..........

6. Name a human disease caused by viruses ..........

7. Name an animal disease caused by viruses ..........

8. Name a plant disease caused by viruses ..........

9. Give the cause, effects, transmission, prevention and treatment of the disease called AIDS.

   Cause ..........

   ..........

   Effects ..........

   ..........

Transmission ..............................................................................

..............................................................................

Prevention ..............................................................................

..............................................................................

Treatment ..............................................................................

..............................................................................

10. Why is it difficult to classify viruses as living things? ..............................................................................

..............................................................................

11. Give one helpful and one harmful effect of viruses.

Helpful ..............................................................................

Harmful ..............................................................................

12. Describe, with the help of diagrams, replication in viruses.

# Unit 3

# Defence System in Humans

1. A pathogen is ..........

2. Name two ways the human body is able to protect itself against pathogens.

   a) ..........

   b) ..........

3. Name four barriers the general defence system uses to prevent pathogens entering the body.

   a) ..........

   b) ..........

   c) ..........

   d) ..........

4. The specific defence system protects the body against diseases by making ..........

   ..........

   ..........

   ..........

5. Immunity is the body's ability to ..........

   ..........

   ..........

6. Name three organs in the body involved in immunity.

   a) ..........

   b) ..........

   c) ..........

7. What is an antigen? ..........

   ..........

   ..........

   ..........

8. What is an antibody? ..........

9. What is a vaccine? ..........

10. Describe how the body gets natural active immunity ..........

11. What is the function of a phagocytic white blood cell? ..........

12. Where are antibodies made? ..........

# Unit 3

# Plant Reproduction

1. Describe the two types of reproduction in plants.

   a) Sexual reproduction ..........................................................................

   ..........................................................................

   b) Asexual reproduction ..........................................................................

   ..........................................................................

2. Give one example of asexual reproduction in plants ..........................................

3. What is the function of the flower? ..........................................

4. Name the parts of the flower labelled in the diagram.

   A..........................................

   B..........................................

   C..........................................

   D..........................................

   E.......................................... F..........................................

   G.......................................... H..........................................

5. Give one function for each flower part.

| Flower part | Function |
|---|---|
| Sepal | |
| Petal | |
| Stamen | |
| Anther | |
| Pollen | |
| Carpel | |
| Ovary | |
| Ovule | |

6. Where is pollen produced in the flower? ..........

7. The diagram shows a grass flower. Name part A and state what is produced in it?

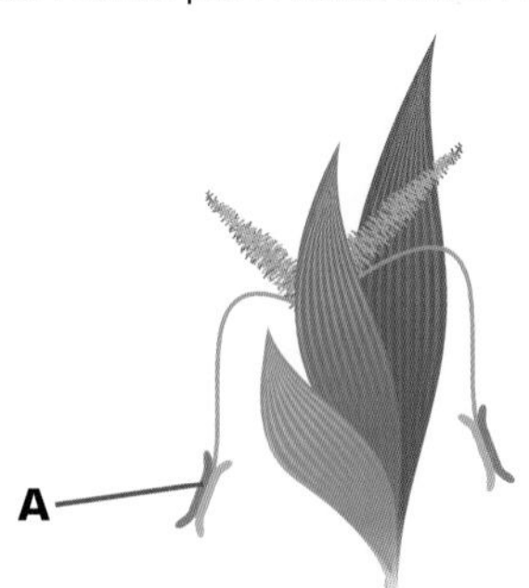


A.......... ..........

How is this flower pollinated? ..........

List two features that show how this flower is adapted to this type of pollination.

a) ..........

b) ..........

8. Name the two nuclei in the pollen grain diagram.

..........

..........

9. The female egg cell and polar nuclei are made in ..........

which is made inside the ..........

10. Name the structure shown in the diagram below ..........

Where would you find this structure? ..........

Name the parts of the diagram.

..........

..........

11. Pollination is the ..........

12. Define the following terms:

Self-pollination ..........

..........

Cross-pollination ..........

..........

13. Name two methods of cross-pollination.

a) ..........

b) ..........

14. Describe two differences between wind-pollinated and insect-pollinated flowers.

Give an example of each method.

| Wind-pollinated flowers | Insect-pollinated flowers |
|---|---|
| 1. | 1. |
| | |
| 2. | 2. |
| | |
| Example: | Example: |
| | |

15. What is meant by fertilisation? ..........

..........

..........

16. In the flowering plant there is a double fertilisation. Complete the following sentences:

Male nucleus + egg nucleus → ..........

Male nucleus + 2 polar nuclei → ..........

17. Name the two main types of seeds.

a) ..........

b) ..........

18. Name the parts of a broad bean seed.

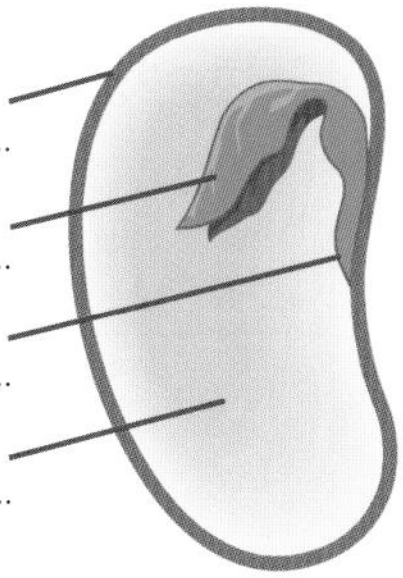

..........................................................

..........................................................

..........................................................

..........................................................

19. Give one function for each seed part.

Testa ..........................................................

Radicle ..........................................................

Plumule ..........................................................

Cotyledon ..........................................................

20. Some flowers have nectaries. How are these flowers pollinated? ..........................................................

Explain your answer ..........................................................

..........................................................

..........................................................

..........................................................

21. In which part of the flower does the seed form? ..........................................................

22. Name the part of the flower that may develop into a fruit ..........................................................

23. Name the male structure in the flower ..........................................................

24. Name the female structure in the flower ..........................................................

25. Seedless fruit can be produced using ..........................................................

..........................................................

..........................................................

26. Give one reason why seeds are scattered away from the parent plant ..........................................................

..........................................................

..........................................................

..........................................................

27. Name four ways seeds are dispersed (scattered) and give two examples for each method.

| Method | Examples |
|---|---|
| | |
| | |
| | |
| | |
| | |
| | |
| | |
| | |

28. State the method of dispersal for each of the seeds shown below.

A..........................................................................................................

B..........................................................................................................

C..........................................................................................................

D..........................................................................................................

E..........................................................................................................

F..........................................................................................................

G..........................................................................................................

A B C D E F G

29. After the seeds are scattered they have a resting period called ..........................................................

30. Give two benefits of the resting period for seeds.

a)..........................................................................................................

b)..........................................................................................................

31. **brightly coloured petals, feathery stigmas, anthers within petals, anthers outside petals, nectaries, petals reduced or absent.**

From the above list, choose three features/characteristics for each pollination method.

| An insect-pollinated flower | A wind-pollinated flower |
|---|---|
| | |
| | |
| | |

32. What is the endosperm? ..........

..........

33. Describe how the endosperm is formed ..........

..........

..........

34. Name the stage seeds go through prior to germination..........

..........

Give one advantage of this stage to the plant ..........

..........

..........

35. Give three examples of how fruits are involved in seed dispersal.

a)..........

b)..........

c)..........

# Unit 3

# Germination

1. Germination is ........................................

2. Name three conditions needed for seeds to germinate.

   a) ........................................

   b) ........................................

   c) ........................................

3. When a seed begins to germinate it takes in ........................................

   to activate the ........................................

4. During the early part of germination ........................................

   break down the food stored in the seed into a simple form.

5. Name the enzyme that acts on each food type and the end products.

| Food type | Enzyme | End product(s) | Use made of food |
|---|---|---|---|
| Starch | | | |
| | | | |
| Lipid | | | |
| | | | |
| Protein | | | |
| | | | |

6. List the stages of germination ........................................

........................................

........................................

........................................

........................................

........................................

7. In which of the following test tubes did the seeds germinate? ................................................

   Why? ................................................

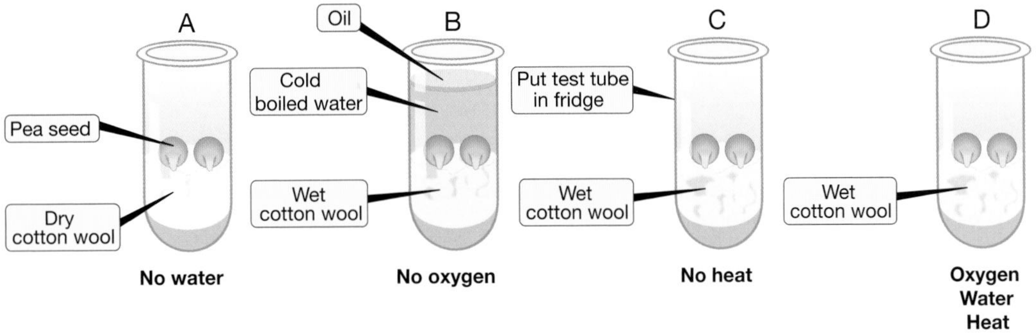

A B C D

Oil

Cold boiled water

Pea seed

Dry cotton wool

Wet cotton wool

Put test tube in fridge

Wet cotton wool

Wet cotton wool

No water

No oxygen

No heat

Oxygen
Water
Heat

8. In the experiment to show digestive activity during germination, the diagram below shows the result when the plates were flooded with iodine and then drained off.

   Why did clear areas form around the raw seeds, but not around the boiled seeds? ................................................

   ................................................

   ................................................

   Why were the seeds boiled? ................................................

   ................................................

'Raw' seeds

'Boiled' seeds

9. The following are some of the stages involved in sexual reproduction in the flowering plant:

   **germination, fertilisation, seed and fruit formation, pollination, dispersal**

   Write the stages in the order they occur ................................................

   ................................................

10. Distinguish clearly between pollination and fertilisation ................................................

    ................................................

    ................................................

11. State a location in the seed where food is stored ..........

12. What is meant by the term 'digestion'? ..........

..........

Why does digestion occur in seeds during germination? ..........

..........

13. Answer the following questions in relation to practical work you carried out to investigate digestive activity in germinating seeds.

a) Name a plant that provides suitable seeds for this investigation ..........

..........

b) The seeds were divided into two batches. One batch was used untreated.

How did you treat the other batch of seeds before using them in the investigation? ..........

..........

Explain why you treated the second batch of seeds in the way described in (b) ..........

..........

..........

..........

# Unit 3

# Vegetative Reproduction

1. What is vegetative propagation? ..........

2. Give one difference between sexual reproduction and asexual reproduction ..........

..........

..........

..........

3. Vegetative propagation is a form of ..........reproduction.

4. Asexual propagation does not involve ..........

5. Name four methods of natural and artificial vegetative reproduction and give examples.

| **Natural** | **Example** | **Artificial** | **Example** |
|---|---|---|---|
| | | | |
| | | | |
| | | | |
| | | | |

6. Describe, using diagrams, how one artificial method is carried out by gardeners.

7. The letters show the stages in the process of tissue culturing. However, they are not in the correct order.

   A. Cells grow rapidly into small masses of tissue.

   B. The plant material is transferred to plates containing sterile nutrient agar jelly.

   C. The tiny plantlets are transferred into potting trays where they develop into plants.

   D. Plant hormones are added to stimulate the cells to divide.

   E. More growth hormones are added to stimulate the growth of roots and stems.

   F. Small amounts of parent tissue or a number of cells are taken.

   Re-write the letters in the correct order ..........................................................

8. Give two advantages and two disadvantages of sexual and asexual reproduction in plants.

| Sexual reproduction | Asexual reproduction |
|---|---|
| **Advantages** | |
| 1. | 1. |
| | |
| 2. | 2. |
| | |
| **Disadvantages** | |
| 1. | 1. |
| | |
| 2. | 2. |
| | |

# Unit 3

# Human Reproduction

1. Humans reproduce as a result of ................................ reproduction.

2. Give two functions of the male reproductive organs.

   a) ................................

   b) ................................

3. Give one function for each of the following:

   Testes (singular – testis) ................................

   Scrotum ................................

   Epididymis ................................

   Sperm tube ................................

   Glands ................................

   Urethra ................................

   Penis ................................

4. Name the labelled parts of the male reproductive system.

   A................................

   B................................

   C................................

   D................................

   E................................

   F................................

   G................................

   H................................

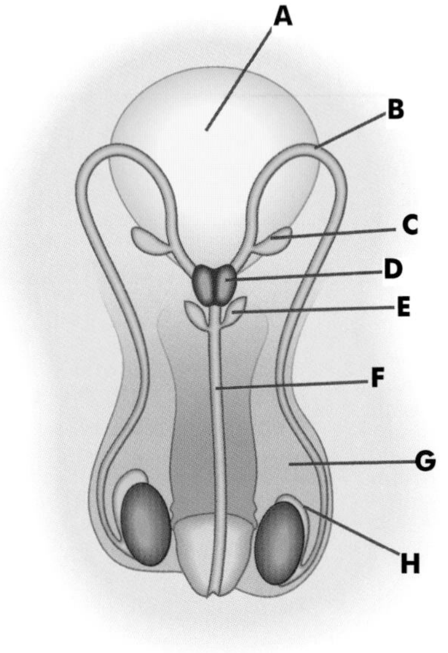

A.......................................

B.......................................

C.......................................

D.......................................

E.......................................

F.......................................

G.......................................

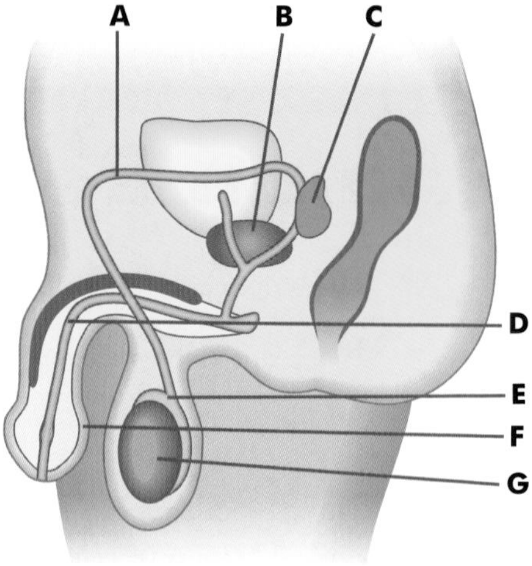

5. Give two functions of the male hormone testosterone.

a).......................................

b).......................................

6. Give one possible cause and treatment of male infertility.

Cause .......................................

Treatment .......................................

7. Give two functions of the female reproductive organs.

a).......................................

b).......................................

8. Name the labelled parts of the female reproductive system

A.......................................

B.......................................

C.......................................

D.......................................

E.......................................

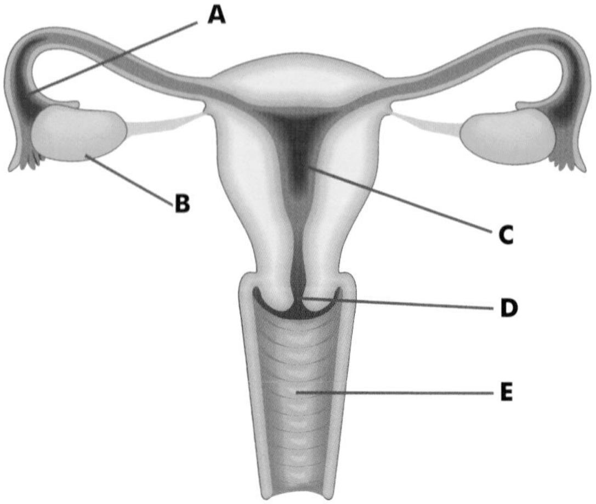

A .............................................................................................. A

B .............................................................................................. B

C .............................................................................................. C

D .............................................................................................. D

E .............................................................................................. E

Anus

9. Give one function for each of the following:

   Ovary ..............................................................................................

   ..............................................................................................

   Fallopian tube (oviduct) ..............................................................................................

   ..............................................................................................

   Womb (uterus) ..............................................................................................

   ..............................................................................................

   Vagina ..............................................................................................

   ..............................................................................................

10. List two functions of the female hormones oestrogen and progesterone.

    a) ..............................................................................................

    b) ..............................................................................................

11. What is the menstrual cycle in females? ..............................................................................................

    ..............................................................................................

12. The menstrual cycle is controlled by ..............................................................................................

13. The menstrual cycle usually takes place over .................................. but can be longer or shorter.

14. What is meant by ovulation? ..............................................................................................

    ..............................................................................................

15. Around what time in the cycle does ovulation usually take place? ..............................................................................................

    ..............................................................................................

16. What happens during menstruation? ..........................................

..........................................

17. When does menstruation take place in the cycle? ..........................................

18. What is meant by the fertile period? ..........................................

..........................................

19. When in the cycle is the fertile period? ..........................................

20. What is the function of FSH in the menstrual cycle? ..........................................

..........................................

..........................................

21. Give two effects of oestrogen in the menstrual cycle.

    a) ..........................................

    b) ..........................................

22. Give two effects of progesterone in the menstrual cycle.

    a) ..........................................

    b) ..........................................

23. Give an effect of LH in the menstrual cycle ..........................................

..........................................

..........................................

24. Give one cause and one treatment for female infertility.

    Cause ..........................................

    Treatment ..........................................

25. Where does the penis release the sperm during intercourse? ..........................................

26. What is meant by fertilisation? ..........................................

..........................................

    Where does it take place? ..........................................

27. What is meant by in-vitro fertilisation? ..........................................

..........................................

28. What is the placenta? ..........

..........

..........

29. What is the function of the placenta? ..........

..........

..........

30. What is meant by implantation? ..........

..........

..........

31. Where does lactation occur? ..........

32. Lactation is controlled by ..........

33. Give two benefits of breastfeeding.

a) ..........

b) ..........

34. The diagram shows the baby in the womb. Name the parts A, B and C.

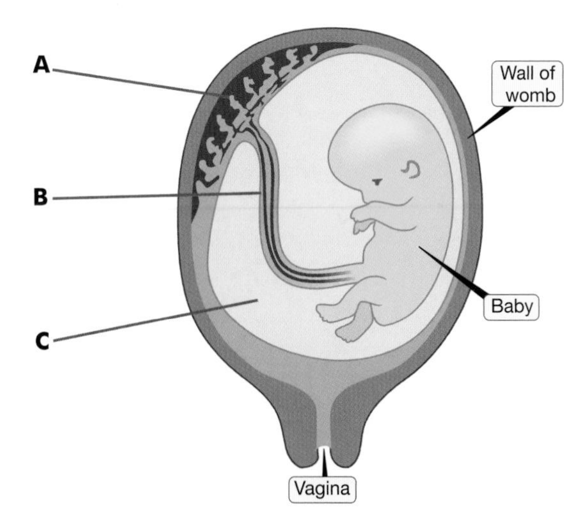


A ..........

B ..........

C ..........

35. In the placenta substances pass between the mother's blood and the baby's blood.

Which way do the following substances pass? Tick the correct box in each row.

| Substance | Mother's blood to baby's blood | Baby's blood to mother's blood |
|---|---|---|
| Poisons from cigarette smoke | | |
| Oxygen | | |
| Digested food | | |
| HIV virus | | |

36. What is meant by pregnancy? ........................................................................................................

........................................................................................................................................................

Approximately how long does pregnancy take in humans? ..............................................................

37. What is meant by contraception? ................................................................................................

........................................................................................................................................................

38. Give one natural, chemical, surgical and mechanical method of contraception.

Natural ..............................................................................................................................................

Chemical ............................................................................................................................................

Surgical ..............................................................................................................................................

Mechanical .........................................................................................................................................